Discourse on Method *and* Meditations on First Philosophy

Discourse on Method *and* Meditations on First Philosophy

third edition

René Descartes

Translated by
Donald A. Cress

Hackett Publishing Company
Indianapolis/Cambridge

René Descartes: 1596–1650

Discourse on Method was originally published in 1637.
Meditations on First Philosophy was originally published in 1641.

99 98 97 96 95 94 2 3 4 5 6 7

Printed in the United States of America

Cover design by Listenberger Design & Associates
Interior design by Dan Kirklin

For further information please address

Hackett Publishing Company, Inc.
P.O. Box 44937
Indianapolis, Indiana 46244-0937

Library of Congress Cataloging in Publication Data

Descartes, René, 1596–1650.
[Discours de la méthode, English]
Discourse on method; and, Meditations on first philosophy/René Descartes; translated by Donald A. Cress.—3rd ed.
p. cm.
ISBN 0-87220-173-2. ISBN 0-87220-172-4 (pbk.)
1. Methodology. 2. Science—Methodology. 3. First philosophy. I. Descartes, René, 1596–1650. Meditationes de prima philosophia. English. 1993. II. Cress, Donald A. III. Title.
B1848.E5C73 1993
194—dc20 93-18772
CIP

The paper used in this publication meets the minimum requirements of American National Standard for Information Sciences—Permanence of Paper for Printed Library Materials, ANSI Z39.48-1984.

∞

CONTENTS

DISCOURSE ON METHOD

MEDITATIONS ON FIRST PHILOSOPHY

EDITOR'S PREFACE

René Descartes was born March 31, 1596, in a small town in Touraine called La Haye (now called La Haye-Descartes or simply Descartes). When he was about ten years old, his father sent him to the Collège Henri IV at La Flèche, a newly formed school which was soon to become the showcase of Jesuit education and one of the outstanding centers for academic training in Europe. Later in his life Descartes looked with pride on the classical education he received from the Jesuits, even though he did not always find agreeable what the Jesuits taught him. He especially found the scholastic Aristotelianism taught there distasteful, although he did cherish his training in many other disciplines—particularly mathematics.

Descartes left La Flèche in 1614 to study civil and canon law at Poitiers, and by 1616 had received the baccalaureate and licentiate degrees in law. In 1618 Descartes joined the army of Prince Maurice of Nassau as an unpaid volunteer, but apparently he never saw combat. He seems to have been more interested in using military service as a means of seeing the world.

During a tour of duty in Germany, events of lifelong importance happened to Descartes. In November of 1619 he was sitting in a *poêle,* a small stove-heated room, meditating on the disunity and uncertainty of his knowledge. He marveled at mathematics, a science in which he found certainty, necessity, and precision. How could he find a basis for all knowledge so that it might have the same unity and certainty as mathematics? Then, in a blinding flash, Descartes saw the method to be pursued for putting all the sciences, all knowledge, on a firm footing. This method made clear both how new knowledge was to be achieved and how all previous knowledge could be certain and unified. That evening Descartes had a series of dreams that seemed to put a divine stamp of approval on his project. Shortly thereafter Descartes left military service.

Throughout the early part of his life, Descartes was plagued by a sense of impotence and frustration about the task he had set about to accomplish: a new and stable basis for all knowledge. He had the programmatic vision, but he seemed to despair of being able to work it out in detail. Thus, perhaps we have an explanation for the fact that Descartes, during much of the 1620s, threw himself into the pursuit of the good life. Travel, gambling, and dueling seemed especially to attract his attention.

This way of life ended in 1628, when, through the encouragement of

Cardinal de Bérulle, Descartes decided to see his program through to completion. He left France to avoid the glamour and the social life; he renounced the distractions in which he could easily lose himself and forget what he knew to be his true calling. He departed for Holland, where he would live for the next twenty years.

It was during this period that Descartes began his *Rules for the Direction of the Mind* and wrote a short treatise on metaphysics, although the former was not published during his lifetime and the latter seems to have been destroyed by him. Much of the early 1630s was taken up with scientific questions. However, Descartes's publication plans were abruptly altered when he learned of the trial of Galileo in Rome. Descartes decided, as Aristotle had centuries before, that philosophy would not be sinned against twice. He suppressed his scientific treatise, *The World or Treatise on Light.*

In 1637 Descartes published in French a *Discourse on the Method for Conducting One's Reason Rightly and for Searching for Truth in the Sciences;* it introduced three treatises which were to exemplify the new method: one on optics, one on geometry, and one on meteorology. Part IV of the introductory *Discourse* contained, in somewhat sketchy form, much of the philosophical basis for constructing the new system of knowledge.

In response to queries about this section, Descartes prepared a much lengthier discussion of the philosophical underpinnings for his vision of a unified and certain body of human knowledge. This response was to be his *Meditations on First Philosophy,* completed in the spring of 1640—but not published until August, 1641. Attached to the *Meditations* were sets of objections and queries sent by readers who had read the manuscript, plus Descartes's replies to each set.

The period following the publication of the *Meditations* was marked by controversy and polemics. Aristotelians, both Catholic and Protestant, were outraged; many who did not understand Descartes's teachings took him to be an atheist and a libertine. In spite of all of this clamor, Descartes hoped that his teachings would replace those of Aristotle. To this end he published in 1644 his *Principles of Philosophy,* a four-part treatise which he hoped would supplant the Aristotelian scholastic manuals used in most universities. The last important work to be published during his lifetime was his *Passions of the Soul,* in which Descartes explored such topics as the relationship of the soul to the body, the nature of emotion, and the role of the will in controlling the emotions.

In 1649 Queen Christina of Sweden convinced Descartes that he should come to Stockholm in order to teach her philosophy. Christina seems to have regarded Descartes more as a court ornament for her amusement and edification than as a serious philosopher; however, it was the brutal winter of 1649 that proved to be Descartes's undoing. Of the climate in Sweden Descartes was to say: "It seems to me that men's thoughts freeze here during winter, just as does the water." Descartes caught pneumonia early in February of 1650 and, after more than a week of suffering, died on February 11.

SELECTED BIBLIOGRAPHY

A. STANDARD EDITION

Oeuvres de Descartes, publiés par Charles Adam et Paul Tannery, 13 volumes. Paris: Cerf, 1897–1913. (Vols. 1–11 contain Descartes's writings; vol. 12 contains Charles Adam's *Vie et oeuvres de Descartes*; vol. 13 is a supplementary volume containing correspondence, biographical material, and various indexes.) It has been updated (Paris: Vrin, 1964ff.), and additional correspondence has been appended to various volumes. More accurate identifications of dates and addressees have been supplied; especially important is the inclusion of Descartes's correspondence with Huygens. This edition is commonly cited as AT.

B. ENGLISH TRANSLATIONS

The Philosophical Works of Descartes. 2 volumes. Rendered into English by Elizabeth S. Haldane and G.R.T. Ross. 2nd edition, corrected. Cambridge: Cambridge University Press, 1931.

Until 1984 this often reprinted but error-plagued set of volumes was the standard translation of many of Descartes's central works. Virtually all twentieth-century Anglo-American scholars made use of Haldane-Ross. This edition was commonly cited as HR.

The Philosophical Writings of Descartes. 3 vols. Translated by John Cottingham, Robert Stoothoff, Dugald Murdoch, and Anthony Kenny. Cambridge: Cambridge University Press, 1984, 1991.

This translation is a welcome replacement of HR. The first volume contains philosophical works other than those related to the *Meditations*; the second volume contains the *Meditations* and the *Replies to Objections*; the third volume contains Descartes's philosophical correspondence and much of the *Conversation with Burman*. This edition is commonly cited as CSM.

Descartes, Philosophical Letters, Edited and translated by Anthony Kenny. Oxford: Oxford University Press, 1970; reprinted Minneapolis: University of Minnesota, 1981.

Descartes's correspondence is an invaluable resource that complements his published works. For twenty years this was the standard English translation of Descartes's philosophical correspondence. Although the translations are reliable, references in the footnotes and the index should be used with care, as there are many errors in the Oxford edition, and they were not corrected in the later reprint. This volume was commonly cited as K. It has now been incorporated into volume three of CSM; errors have been corrected, and additional correspondence has been included.

Descartes' Conversation with Burman. Translated, with notes, by John Cottingham. Oxford: Oxford University Press, 1976.

Housed in the Library of the University of Göttingen is a manuscript that purports to chronicle a discussion between Descartes and the young Dutch theologian Francis Burman. Burman had chosen several texts from Descartes's writings for discussion. Sometimes he would criticize the doctrine in the text; sometimes he would simply ask for clarification. Descartes's (?) replies are always interesting and nearly always shed light on difficult passages in his published works. Cottingham's extensive commentary is both interesting and helpful. It is commonly cited as CB. Since volume three of CSM does not provide the complete text of the *Conversation with Burman*, this translation must continue to be consulted.

Discourse on Method, Optics, Geometry, and Meteorology. Translated by Paul J. Olscamp. Indianapolis: Bobbs-Merrill, 1965.

Treatise on Man. French text with translation and commentary by Thomas Steele Hall. Cambridge: Harvard University Press, 1972.

Descartes: Le Monde, ou Traité de la lumière. Translation and introduction by Michael Sean Mahoney. New York: Abaris Books. 1979.

Descartes: Principles of Philosophy. Translated, with explanatory notes, by Valentine Rodger Miller and Reese P. Miller. Dordrecht: Reidel, 1983.

C. BIBLIOGRAPHIES: 1800–1984

Sebba, Gregor. *Bibliographia Cartesiana: A Critical Guide to the Descartes Literature 1800–1960*. The Hague: Nijoff, 1964.

This is the basic bibliographical tool of pre-1960 Descartes scholarship. It contains a large number of annotations and cross-references; it is well indexed by person and subject matter. Although somewhat weak in its coverage of twentieth-century Anglo-American analytical literature on Descartes, it is outstanding in its coverage of continental scholarship.

Doney, Willis. "Bibliography," in *Descartes: A Collectum of Critical Essays*. New York: Doubleday, 1967, pp. 369–386.

This bibliography largely rectifies Sebba's lack of coverage of pre-1960 analytical works on Descartes. It is concerned chiefly with English titles; it is divided by subject matter.

Chappell, Vere, and Willis Doney. *Twenty-Five Years of Descartes Scholarship, 1960–1984: A Bibliography*. New York: Garland, 1987.

This volume, while neither complete nor adequately indexed, is still the best update of Sebba.

Cress, Donald A. "Canadian and American Dissertations on Descartes and Cartesianism: 1865–1984." *Philosophy Research Archives* 13 (April 1988).

This bibliography provides author, title, university, and location in *Dissertation Abstracts International*, as well as the University Microfilms International order number, when available.

D. BOOKS ON DESCARTES

Beck, Leslie. *The Metaphysics of Descartes*. Oxford: Oxford University Press, 1965.

Butler, R. J., ed. *Cartesian Essays*. Oxford: Basil Blackwell, 1972.

Caton, Hiram. *The Origin of Subjectivity: An Essay on Descartes*. New Haven: Yale University Press, 1973.

Chappell, Vere, ed. *Essays on Early Modern Philosophy from Descartes and Hobbes to Newton and Leibniz*. Vol. I: *René Descartes*. New York: Garland, 1992.

Cottingham, John. *Descartes*. Oxford: Basil Blackwell, 1986.

Cottingham, John, ed. *The Cambridge Companion to Descartes*. New York: Cambridge University Press, 1992.

Cottingham, John. *A Descartes Dictionary*. Oxford and Cambridge, Mass.: Oxford University Press, 1993.

Curley, E. M. *Descartes against the Sceptics*. Cambridge, Mass.: Harvard University Press, 1978.

Dicker, Georges. *Descartes, An Analytical and Historical Introduction*. Oxford: Oxford University Press, 1993.

Doney, Willis, ed. *Descartes, A Collection of Critical Essays*. Garden City: Doubleday, 1967.

Doney, Willis, ed. *Eternal Truths and the Cartesian 'Circle': A Collection of Studies*. New York: Garland, 1987.

Frankfurt, Harry G. *Demons, Dreamers, and Madmen: The Defense of Reason in Descartes's Meditations*. Indianapolis: Bobbs-Merrill, 1970; reprinted New York: Garland, 1987.

Garber, Daniel. *Descartes' Metaphysical Physics*. Chicago: University of Chicago Press, 1992.

Gibson, A. Boyce. *The Philosophy of Descartes*. London: Methuen and Co., 1932; reprinted New York: Garland, 1987.

Gilson, Étienne. *Discours de la méthode: texte et commentaire*. 4th edition. Paris: Vrin, 1967.

Grene, Marjorie. *Descartes*. Minneapolis: University of Minnesota Press, 1985.

Gueroult, Martial. *Descartes' Philosophy Interpreted According to the Order of Reasons*. 2 vols. Translated by Roger Ariew. Minneapolis: University of Minnesota Press, 1984, 1985.

Hooker, Michael, ed. *Descartes, Critical and Interpretive Essays*. Baltimore: Johns Hopkins University Press, 1978.

Judovitz, Dalia. *Subjectivity and Representation in Descartes: The Origins of Modernity*. New York: Cambridge University Press, 1988.

Kenny, Anthony. *Descartes: A Study of His Philosophy*. New York: Random House, 1968; reprinted New York: Garland, 1987.

Markie, Peter J. *Descartes's Gambit*. Ithaca: Cornell University Press, 1986.

Matthews, Gareth B. *Thought's Ego in Augustine and Descartes*. Ithaca: Cornell University Press, 1992.

Moyal, Georges J. D., ed. *René Descartes: Critical Assessments*. 4 vols. New York: Routledge, 1991.

Rorty, Amélie Oksenberg, ed. *Essays on Descartes' Meditations*. Berkeley: University of California Press, 1986.

Sesonske, Alexander, and Noel Fleming, eds. *Meta-Meditations: Studies in Descartes*. Belmont: Wadsworth Publishing Co., 1965.

Sorell, Tom. *Descartes*. Oxford: Oxford University Press, 1987.

Tweyman, Stanley, ed. *René Descartes: Meditations on First Philosophy in Focus*. New York: Routledge, 1993.

Verbeek, Theo. *Early Reactions to Cartesian Philosophy: 1637–1650*. Carbondale: Southern Illinois University Press, 1992.

Voss, Stephen, ed. *Essays on the Philosophy and Science of René Descartes*. Oxford: Oxford University Press, 1992.

Watson, Richard A. *The Breakdown of Cartesian Metaphysics*. Atlantic Highlands, N.J.: Humanities Press International, 1987.

Williams, Bernard. *Descartes, the Project of Pure Enquiry*. London: Penguin Books, 1978.

Wilson, Margaret. *Descartes*. London: Routledge and Kegan Paul, 1978.

Wolf-Devine, Celia. *Descartes on Seeing: Epistemology and Visual Perception*. Carbondale: Southern Illinois University Press, 1993.

Discourse
on
the Method for
Rightly Conducting
One's Reason
and
for Seeking Truth in
the Sciences

NOTE ON THE TRANSLATION

The translation is based on the original French version (1637) of the *Discourse on Method* found in volume six of the Adam and Tannery edition of Descartes's works (Paris: Vrin, 1965). The numbers in the margins of this translation refer to the pagination of the Adam and Tannery edition.

D.A.C.

DISCOURSE ON THE METHOD FOR RIGHTLY CONDUCTING ONE'S REASON AND FOR SEEKING TRUTH IN THE SCIENCES

If this discourse seems too long to be read at one sitting, one might split it into six parts. In the first, one will find various discussions concerning the sciences. In the second part, the chief rules of the method which the author has been seeking. In the third part, some of the rules of morality which the author has derived from this method. In the fourth part, the reasons by which the author proves the existence of God and of the human soul, which are the foundations of his metaphysics. In the fifth part, the order of the questions in physics which the author has sought—and particularly the explanation of the heart's movement and other difficulties which pertain to medicine, as well as the difference between our soul and that of animals. And in the final part, what things the author believes are required to advance further in the study of nature than the author has done, and what reasons moved him to write.

PART ONE

Good sense is the most evenly distributed commodity in the world, for each of us considers himself to be so well endowed therewith that even those who are the most difficult to please in all other matters are not wont *2* to desire more of it than they have. It is not likely that anyone is mistaken about this fact. Rather, it provides evidence that the power of judging rightly and of distinguishing the true from the false (which, properly speaking, is what people call good sense or reason) is naturally equal in all men. Thus the diversity of our opinions does not arise from the fact that some people are more reasonable than others, but simply from the fact that we conduct our thoughts along different lines and do not consider the same things. For it is not enough to have a good mind; the main thing is to use it well. The greatest souls are capable of the greatest vices as well as of the greatest virtues. And if they always follow the correct path, those who move forward only very slowly can make much greater progress than do those who run and stray from it.

For myself, I have never presumed that my mind was in any respect more perfect than anyone else's. In fact, I have often longed to have as quick a wit or as precise and distinct an imagination or as full and responsive a memory as certain other people. And I know of no other qualities that aid in the perfection of the mind. For as to reason or good sense, given that it alone makes us men and distinguishes us from animals, I prefer to believe that it exists whole and entire in each one of us. In this belief I am following the standard opinion held by philosophers who say 3 that there are differences of degree only among accidents, but not among forms or natures of individuals of the same species.

But I shall have no fear of declaring that I think I have been fortunate; I have, since my youth, found myself on paths that have led me to certain considerations and maxims from which I have formed a method by means of which, it seems to me, I have the ways to increase my knowledge by degrees and to raise it gradually to the highest point to which the mediocrity of my mind and the short span of my life can allow it to attain. For I have already reaped from it such a harvest that, though as regards judgments I make of myself, I try always to lean toward caution, rather than toward presumption, and though, looking with a philosopher's eye at the various actions and enterprises of men, there is hardly one that does not seem to me vain and useless, I always take immense satisfaction in the progress that I think I have made in the search for truth; and I envisage such hopes for the future that if, among the occupations of men, as men, there is one which may be solidly good and important, I dare believe that it is the occupation I have chosen.

All the same, it could be that I am mistaken; and what I have taken for gold and diamonds may perhaps be nothing but copper and glass. I know how much we are prone to be mistaken in those things that deeply affect us, and also how judgments made by our friends must be held suspect when these judgments are in our favor. But I would be very happy to show 4 in this discourse the paths that I have followed and to present my life as if in a picture, so that each person may judge it; learning what people commonly think about it may be a new means of teaching myself, which I shall add to those that I am accustomed to employing.

Thus my purpose here is not to teach the method that everyone ought to follow in order to conduct his reason correctly, but merely to show how I have tried to conduct mine. Those who take it upon themselves to give precepts ought to regard themselves as more competent than those to whom they give them; and if they are found wanting in the least detail, they are blameworthy. But, putting forward this essay as merely a history—or, if you prefer, a fable—in which, among the examples one can imitate, one also finds perhaps several others which one is right in not following,

I hope that the essay will be useful to some, while harmful to none, and that my openness will be to everyone's liking.

I have been raised on letters from my childhood, and because I was convinced that through them one might acquire a clear and steady knowledge of everything that is useful for life, I possessed a tremendous desire to learn them. But, as soon as I completed this entire course of study, at the end of which one is ordinarily received into the ranks of the learned, I changed my mind entirely. For I was embarrassed by so many doubts and errors, which appeared in no way to profit me in my attempt at learning, except that more and more I discovered my ignorance. And nevertheless, I was in one of the most celebrated schools of Europe, where [5] I thought there ought to be learned men—if in fact there were any such men in the world. I learned everything the others learned; and, not judging the disciplines taught there to be enough, I even went through every book I could lay my hands on that treated those disciplines considered the most curious and unusual. Moreover, I knew what judgments others were making about me; and it was apparent to me that I was rated no less than my peers, even though there already were those among them who were destined to succeed our teachers. And finally our century seems to me just as flourishing and as fertile in good minds as any of the preceding ones. All of this caused me to feel free to judge everybody else by myself, and to think that there has been no body of knowledge in the world which was of the sort that I had previously hoped to find.

Yet I never ceased admiring the academic exercises with which we occupied ourselves in school. I realized that the languages one learns there are necessary for the understanding of classical texts; that the gracefulness of fables awakens the mind; that memorable deeds recounted in histories uplift it, and, if read with discretion, aid in forming one's judgment; that reading good books is like a conversation with the noblest people of past centuries—their authors—indeed, even a studied conversation in which they uncover only the best of their thoughts; that eloquence has incomparable power and beauty; that poetry has a ravishing delicacy and sweetness; [6] that mathematics contains very subtle inventions that can serve as much to satisfy the curious as to facilitate the arts and to diminish men's labour; that writings dealing with morals contain many lessons and exhortations to virtue that are quite useful; that theology teaches one how to go to heaven; that philosophy provides the means of speaking with probability about all things and of being held in admiration by the less learned; that law, medicine, and the other sciences bestow honors and riches upon those who cultivate them. And thus it is good to have examined all of these disciplines, even the most superstition-ridden and false of them, so that one might know their true worth and guard against being deceived.

But I thought that I had already given enough time to languages and also even to the reading of ancient books—to their histories and to their fables. For it is about the same to converse with those of other centuries as it is to travel. It is good to know something of the customs of various peoples, in order to judge our own more soundly and not to think that everything that is contrary to our way of doing things is worthy of scorn and against reason, as those who have seen nothing commonly think. But when one takes too much time traveling, one becomes finally a stranger in one's own country; and when one is too curious about things that took place in past centuries, one ordinarily remains quite ignorant of what is taking place in one's own century. Moreover, fables make one imagine *7* many events to be possible which really are impossible. And even the most accurate histories, if they neither alter nor augment the significance of things, in order to render them more worthy of being read, at least almost always omit the basest and least illustrious details, and thus the remainder does not appear as it really is, and those who govern their own conduct on the basis of the examples drawn from it are subject to falling into the extravagances of the knights of our novels and to conceiving plans that are beyond their powers.

I held eloquence in high regard and I loved poetry, but I believed that they were both gifts of the mind—not fruits of study. Those who possess the most forceful power of reasoning and who best order their thoughts so as to render them clear and intelligible can always best persuade one of what they are proposing, even if they speak only the dialect of Lower Brittany and have never learned rhetoric.[1] And those who are in possession of the most pleasing rhetorical devices and who know how to express them with the greatest of embellishment and sweetness will not fail to be the greatest poets, even if the art of poetry be unknown to them.

I took especially great pleasure in mathematics because of the certainty and the evidence of its arguments. But I did not yet notice its true usefulness and, thinking that it seemed useful only to the mechanical arts, I was astonished that, because its foundations were so solid and firm, no one had built anything more noble upon them. On the other hand, I *8* compared the writings of the ancient pagans who discuss morals to very proud and magnificent palaces that are built on nothing but sand and

1. This dialect was considered rather barbarous and hardly suitable for sophisticated literary endeavors.

mud. They place virtues on a high plateau and make them appear to be valued more than anything else in the world, but they do not sufficiently instruct us about how to know them; and often what they call by such a fine-sounding name is nothing more than insensibility, pride, despair, or parricide.

I revered our theology, and I desired as much as the next man to go to heaven; but having learned as something very certain that the road is no less open to the most ignorant than to the most learned, and that the revealed truths leading to it are beyond our understanding, I would not have dared to subject them to my feeble reasonings. And I believed that, in order to undertake the examination of these truths and to succeed in doing so, it was necessary to have some extraordinary assistance from heaven and to be more than a man.

Of philosophy I shall say only that, aware that philosophy has been cultivated over several centuries by the most excellent minds who have ever lived and that, nevertheless, there is nothing about which there is not some dispute—and thus nothing that is not doubtful—I was not so presumptuous as to hope to fare any better than the others; and that, considering how there can be various opinions that are held by learned people about the very same matter without there ever being any more than one opinion being true, I took to be virtually false everything that was merely probable.

As to the other sciences, since they derive their principles from philosophy, I judged that one could not have built anything solid upon foundations *9* having so little firmness. And neither the honor nor the monetary gain they promised was sufficient to incite me to learn them, for I did not see myself, thank God, as being in a condition that forced me to make a trade out of knowledge for the enhancement of my fortune. And although I did not make a point of rejecting glory in the manner of the cynic, still I made light of that glory that was acquired only through false pretenses. And finally, as to the evil doctrines, I believed I already knew them for what they were worth, so as not to be subject to being deceived either by the promises of an alchemist, by the predictions of an astrologer, by the ruses of a magician, or by the artifices or boasting of anybody who makes a point of claiming to have more knowledge than he actually has.

That is why, as soon as age permitted me to escape the tutelage of my teachers, I left the study of letters completely. And resolving to search for no other knowledge than what I could find within myself, or in the great book of the world, I spent the rest of my youth traveling, seeing various courts and armies, frequenting peoples of varied humors and conditions, gathering varied experiences, testing myself in the encounters which fortune sent my way, and everywhere so reflecting upon what came my way that I could draw some profit from it. For it seemed to me that I could discover much more truth in the reasonings that each person makes concerning matters that are important to him, whose outcome ought to *10* cost him dearly later on if he has judged incorrectly, than in those reasonings that a man of letters makes in his private room, which touch on speculations producing no effect, and which for him have no other consequence except perhaps that the more they are removed from common sense, he will derive all the more vanity from them, for he will have to employ that much more wit and artifice in attempting to make them probable. And I have always had an especially great craving for learning to distinguish the true from the false, to see my way clearly in my actions, and to go forward with confidence in this life.

It is true that, while I spent time merely observing the customs of other men, I found hardly anything about which to be confident and that I noticed there was about as much diversity as I had earlier found among the opinions of philosophers. Thus the greatest profit I derived from this was that on realizing that many things, although they seemed very extravagant and ridiculous to us, did not cease being commonly accepted and approved by other great peoples, I learned to believe nothing very firmly concerning what I had been persuaded to believe only by example and custom; and thus I gradually freed myself from many errors that can darken our natural light and render us less able to listen to reason. But after spending many years thus studying in the book of the world and in trying to gain experience, I made up my mind one day also to study myself and to spend all the powers of my mind in choosing the ways which I ought to follow. For me this procedure was much more successful, it *11* seems, than if I had never left either my country or my books.

Part Two

I was in Germany then, where the wars—which are still continuing there[2]—called me; and while I was returning to the army from the coronation of the emperor, the onset of winter held me up in quarters where, finding no conversation with which to be diverted and, fortunately, otherwise having no worries or passions which troubled me, I remained for a

2. The Thirty Years' War (1618–1648).

whole day by myself in a small stove-heated room,[3] where I had complete leisure for communing with my thoughts. Among them, one of the first that I thought of considering was that often there is less perfection in works made of several pieces and in works made by the hands of several masters than in those works on which but one master has worked. Thus one sees that buildings undertaken and completed by a single architect are commonly more beautiful and better ordered than those that several architects have tried to patch up, using old walls that had been built for other purposes. Thus these ancient cities that were once merely straggling villages and have become in the course of time great cities are commonly quite poorly laid out, compared to those well-ordered towns that an engineer lays out on a vacant plain as it suits his fancy. And although, upon considering one by one buildings in the former class of towns, one finds as much art or more than one finds in buildings of the latter class of towns, still, upon seeing how the buildings are arranged—here a large one, there a small one—and how they make the streets crooked and *12* uneven, one will say that it is chance more than the will of some men using their reason that has arranged them thus. And if one considers that there have nevertheless always been officials responsible for seeing that private buildings be made to serve as an ornament for the public, one will know that it is difficult to produce a finely executed product by laboring only on the works of others. Thus I imagined that peoples who, having once been half savages and having been civilized only gradually, have made their laws only to the extent that the inconvenience caused by crimes and quarrels forced them to do so, would not be as well ordered as those who, from the very beginning of their coming together, have followed the fundamental precepts of some prudent legislator. Thus it is quite certain that the state of the true religion, whose ordinances were fixed by God alone, ought to be incomparably better governed than all the others. And, speaking of matters human, I believe that if Sparta flourished greatly in the past, it was not because of the goodness of each of its laws taken by itself, since some of them were very strange and even contrary to good morals, but because, having been invented by only one person, they all tended toward the same goal. And thus I thought that book learning, at least the kind whose arguments are merely probable and have no demonstrations—having been built up from and enlarged gradually by the opinions of

3. There is no need to allege that Descartes sat in or on a stove. *A poêle is* simply a room heated by an earthenware stove. Cf. E. Gilson, *Discours de la méthode: texte et commentaire* (Paris: Vrin, 1976), p. 157.

many different people—does not draw as near to the truth as the simple 13 reasonings that can be made naturally by a man of good sense concerning what he encounters. And thus again I thought that, given the fact that we were all children before being adults and that for a long time it was our lot to be governed by our appetites and our teachers (both were often in conflict with one another, and perhaps none of them consistently gave us the best advice), it is almost impossible for our judgments to be as pure or solid as they would have been had we the full use of our reason from the moment of our birth and had we never been led by anything but our reason.

It is true that one does not see people pulling down all the houses in a city simply to rebuild them some other way and to make the streets more attractive; but one does see that several people do tear down their own houses in order to rebuild them, and that even in some cases they are forced to do so when their houses are in danger of collapsing and the foundations are not very steadfast. Taking this example to heart, I was persuaded that it was not really likely that a single individual might plan to reform a state by changing everything from the very foundations and by toppling it in order to set it up again; nor even also to reform all of the sciences or the order established in the schools for teaching them; but that I could not do better than to try once and for all to get all the beliefs I had accepted from birth out of my mind, so that once I have reconciled them with reason I might again set up either other, better ones or even 14 the same ones. And I firmly believed that by this means I would succeed in conducting my life much better than were I to build only on old foundations or to lean only on the principles of which I permitted myself to be persuaded in my youth without ever having examined whether or not they were true. For although I noticed various difficulties in this operation, still they did not seem irremediable or comparable to those difficulties arising in the reformation of the least things which affect the public. These great bodies are too difficult to raise up once they have been knocked down or even to maintain once they have been shaken; and their falls can only be very violent. Now as to their imperfections, if they have any (and the mere fact of their diversity suffices to assure one that many of them are imperfect), usage has doubtlessly mitigated them and has even imperceptibly averted or corrected a great number of them, for which deliberate foresight could not have provided so well. And finally, these imperfections are almost always more tolerable than what it takes to change them; similarly, the great roads that wind around mountains become gradually so level and suitable by dint of being used frequently, that it is much better to follow these roads than to try to go by a more direct route, climbing over rocks and descending to the bottom of precipices.

This is why I could not approve of all of those trouble-making and quarrelsome types who, called neither by birth nor by fortune to manage public affairs, never cease in their imagination to effect some new reformation. And if I thought there were the slightest thing in this essay by means of which one might suspect me of such folly, I would be very sorry to *15* permit its publication. My plan has never been more than to try to reform my own thoughts and to build upon a foundation which is completely my own. And if, my work having sufficiently pleased me, I show it to you here as a model, it is not for that reason that I wish to advise anyone to imitate it. Perhaps those with whom God has better shared his graces have more lofty plans; but I fear that this plan of mine may already be too arduous for many. The single resolution to detach oneself from all the beliefs one has once accepted as true is not an example that everyone ought to follow; and the world consists almost completely of but two kinds of people and for these two kinds it is not at all suitable: namely those who, believing themselves more capable than they really are, cannot help making premature judgments and do not have enough patience to conduct their thoughts in an orderly manner; thus, if they once take the liberty to doubt the principles they have accepted and to keep away from the common path, they could never keep to the path one must take in order to go in a more forward direction—they would remain lost all of their lives. Now as for those people who have enough reason or modesty to judge that they are less capable to distinguish the true from the false than are others by whom they can be instructed, they ought to content themselves more with following the opinions of these others than to look for better opinions on their own.

And for my part, I would unquestionably have been among these latter *16* persons were I to have had only one master or had never seen the differences that have always existed among the opinions of the most learned people. But having learned since my school days that one cannot imagine anything so strange or unbelievable that it has not been said by some philosopher, and since then, during my travels, having acknowledged that those who have feelings quite contrary to our own are not for that reason barbarians or savages, but that many of them use their reason as much as or more than we do, and having considered how the very same man with his very own mind, having been brought up from infancy among the French or the Germans becomes different from what he would be had he always lived among the Chinese or among cannibals; and how, even to the fashions of our clothing, the same thing that pleased us ten years ago and that perhaps might again please us ten years from now seems to us extravagant and ridiculous. Thus it is more custom and example that persuades us than certain knowledge, and for all that, the majority opinion is not a proof worth anything for truths that are a bit difficult to discover, since it is more likely that only one man has found them than a whole people: I could find no one whose opinions, it seemed to me, ought to be preferred over the others, and I found myself constrained to try to lead myself on my own.

But, like a man who walks alone and in the shadows, I resolved to go *17* so slowly and to use so much circumspection in all things that, if I never advanced but slightly, I would at least avoid falling. Moreover, I did not wish to begin to reject utterly any of these opinions that could have at some time slipped into my head without having been placed there by my reason, until I had already spent sufficient time formulating the outline of the work I was attempting and searching for the true method of arriving at the knowledge of everything my mind was capable of attaining.

In my younger days I studied, among the parts of philosophy, a bit of logic, and in mathematics, some geometrical analysis and algebra—three arts or sciences that seemed destined to contribute something to my plan. But in examining them, I saw that, in the case of logic, its syllogisms and the greater part of its other lessons served more to explain to someone else what one knows, or even, like the art of Lully,[4] to speak without judgment concerning matters about which one is ignorant, than to learn them. And although it contains, in effect, very true and good precepts, nevertheless there are so many others, mixed up with them, that are either harmful or superfluous, that it is almost as hard to separate the latter precepts from the former as it is to draw a Diana or a Minerva from a block of marble that is not yet blocked out. Now, as to the analysis of the ancients and the algebra of the moderns, apart from the fact that they *18* apply only to very abstract matters and seem to have no practical utility, the former is always so subject to the consideration of figures that it cannot exercise the understanding without exhausting the imagination; and in the

4. Ramon Lull (ca. 1236–1315), Catalan philosopher and Franciscan who wrote in defense of Christianity against the Moors by attempting to demonstrate the articles of faith by means of logic. Descartes seems to have encountered a Lullist in Dordrecht who could hold forth on any subject whatever for long periods of time. This encounter, more than any direct contact with the writings of Lull, seems to have colored Descartes's understanding of the "art of Lully." Cf. E. Gilson, *Discours de la méthode: texte et commentaire,* pp. 185–186.

case of algebra, one is so governed by certain laws and symbols that out of it has been made a confused and obscure art that encumbers the mind instead of a science that cultivates it. That is why I believed it necessary to search for another method that, while embracing the advantages of the three, was free from their defects. And since the multiplicity of laws often provides excuses for vices, so that a state is much better when, having but a few laws, its laws are strictly observed; so, in place of the large number of rules of which logic is composed, I believed that the following four rules would be sufficient, provided I made a firm and constant resolution not even once to fail to observe them:

The first was never to accept anything as true that I did not know evidently to be so; that is, carefully to avoid precipitous judgment and prejudice; and to include nothing more in my judgments than what presented itself to my mind with such clarity and distinctness that I would have no occasion to put it in doubt.

The second, to divide each of the difficulties I was examining into as many parts as possible and as is required to solve them best.

The third, to conduct my thoughts in an orderly fashion, commencing with the simplest and easiest to know objects, to rise gradually, as by degrees, to the knowledge of the most composite things, and even supposing an order among those things that do not naturally precede one another. *19*

And last, everywhere to make enumerations so complete and reviews so general that I would be sure of having omitted nothing.

Those long chains of reasoning, each of them simple and easy, that geometricians commonly use to attain their most difficult demonstrations, have given me an occasion for imagining that all the things that can fall within human knowledge follow one another in the same way and that, provided only that one abstain from accepting anything as true that is not true, and that one always maintains the order to be followed in deducing the one from the other, there is nothing so far distant that one cannot finally reach nor so hidden that one cannot discover. And I was not very worried about seeking which of them it would be necessary to begin with; for I already knew that it was with the simplest and easiest to know. And considering that of all those who have already searched for truth in the sciences, only the mathematicians were able to find demonstrations, that is, certain and evident reasons, I did not doubt that it was with these same starting points that they had conducted their examinations; although I expected no further usefulness from them, except that they would accustom my mind to feed upon truths and not to be content with false reasons. But in all of this it was not my plan to try to learn all of the specific sciences that commonly are called mathematics; and seeing that, even though their 20 objects differed, they did not cease to be in accord with one another, in that they consider only the various relations or proportions which are in these objects, I believed it would be more worthwhile were I to examine only these proportions in a general way, and to suppose them to be in concrete objects only to the extent that these objects aid me in making it easier to acquire knowledge of these proportions, and also without in any way binding these proportions to those objects, so that later one can apply them all the better to everything else to which they might pertain. Now, having noticed that, in order to know these proportions, I occasionally needed to consider each of them individually, and sometimes only to remember them, or to gather up several of them together, I believed that, to consider them better in particular, I ought to suppose them as relations between lines, since I found nothing more simple, nothing that I could more distinctly represent to my imagination and my senses; but to remember them or to grasp them all together, I would have had to explicate them by means of certain symbols, the shortest ones possible; and by this means I would borrow all of the better aspects of geometrical analysis and algebra, and I would correct all the defects of the one by means of the other.

In effect, I dare say that the exact observance of these few precepts that I have chosen gave me such a facility for disentangling all the problems to which these two sciences tended, that in the two or three months I spent examining them, having begun from the simplest and most general— 21 and each truth that I found being a rule which I later used to find others— not only did I succeed in several problem areas that I had once judged very difficult, but it also seemed to me toward the end that I could determine, even in those problem areas where I was ignorant, by what means and how far it would be possible to resolve them. In this perhaps I shall not seem to you to be too vain, if you were to consider that, there being but one truth for each thing, anyone who finds it knows as much as one can know about that thing; and that, for example, a child given lessons in arithmetic, having made one addition in accordance with its rules, can be assured of having found everything the human mind can find bearing on the sum he has examined. For thus, the method, which teaches one to follow the true order and to enumerate exactly all the circumstances of what one is seeking, contains everything that gives certainty to the rules of arithmetic.

But what pleased me the most about this method was that by means of it I was assured of using my reason in everything, if not perfectly, then at least as best as I can. Moreover, I felt that in practicing this method my mind was gradually getting into the habit of conceiving its object more rigorously and more distinctly and that, not having subjected it to any particular matter, I promised myself to apply the method just as profitably to the problems of the other sciences, as I had done to problems in algebra. Not that, on account of this, I dared immediately to undertake an examination of whatever presented itself; for even that would have been contrary to the order prescribed by the method. But having noticed that their principles must all be borrowed from philosophy, in which I still *22* found nothing certain, I thought that I ought, above all, to try to establish therein something certain; and I thought that, this being the most important thing in the world, where precipitous judgment and prejudice were most to be feared, I ought not to have tried to succeed at doing so until I had reached a much more mature age than merely twenty-three, which I was then; and I thought that I should previously spend much time preparing myself, as much in rooting out of my mind all the wrong opinions that I had accepted before that time as in accumulating many experiences—later to be the stuff of my reasonings—and in always exercising myself in the method I had prescribed for myself so as to be stronger and stronger in its use.

Part Three

Now just as it is not enough, before beginning to rebuild the house where one lives, to pull it down, to make provisions for materials and architects, or to take a try at architecture for oneself, and also to have carefully worked out the floorplan; one must provide for something else in addition, namely where one can be conveniently sheltered while working on the other building; so too, in order not to remain irresolute in my actions while reason requires me to be so in my judgments, and in order not to cease living during that time as happily as possible, I formulated a provisional code of morals, which consisted of but three or four maxims, that I want to share with you.

The first was to obey the laws and the customs of my country, firmly *23* holding on to the religion in which, by God's grace, I was instructed from childhood, and governing myself in all other things according to the most moderate opinions and those furthest from excess that were commonly accepted in practice by the most sensible of those people with whom I would have to live. For, already beginning to count my own opinions as nothing, since I wished to remit all opinions to examination, I was assured that I could not do better than to follow the opinions of those who were the most sensible. And although there may perhaps be people among the Persians and the Chinese just as sensible as there are among ourselves, it seemed to me that the most useful course of action was to rule myself in accordance with those with whom I had to live, and that, to know their true opinions, I ought to observe what they do rather than what they say, not only because in the corruption of our morals there are few people who are willing to say all they believe, but also because many do not know what they believe; for, given that the action of thought by which one believes something is different from that by which one knows that one believes it, the one often occurs without the other. And among several opinions held equally, I would choose only the most moderate, not only because it is always the most suitable for action and probably the best (every excess usually being bad), but also so as to stray less, in case I am mistaken, from the true road—having chosen one of the two extremes—when it was the other one I should have followed. And in particular I placed among the excesses all of the promises by which one curtails something of one's *24* freedom. Not that I disapprove of laws that, to remedy the inconstancy of weak minds, permit (when one has a good plan or even, for the security of commerce, a plan that is only indifferent) one to make vows or contracts that oblige one to persevere in them; but because I have seen nothing in the world that remains always in the same state and because, for my part, I promised to perfect my judgments more and more, and not to render them worse, I would have believed I committed a grave indiscretion against good sense if, having once approved of something, I obligated myself to take it to be good at a later time when perhaps it would have ceased to be so or when I would have ceased judging it to be good.

My second maxim was to be as firm and resolute in my actions as I could be, and to follow with no less constancy the most doubtful opinions, once I have decided on them, than if they were very certain. In this I would imitate travelers who, finding themselves lost in a forest, ought not wander this way and that, or, what is worse, remain in one place, but ought always walk as straight a line as they can in one direction and not change course for feeble reasons, even if at the outset it was perhaps only chance that made them choose it; for by this means, if they are not going where they *25* wish, they will finally arrive at least somewhere where they probably will be better off than in the middle of a forest. And thus the actions of life often tolerating no delay, it is a very certain truth that, when it is not in our power to discern the truest opinions, we ought to follow the most probable; and even if we observe no more probability in some than in others, nevertheless we ought to fix ourselves on some of them and later consider them no longer as doubtful, insofar as they relate to practical affairs, but as very true and very certain, since reason, which has caused us to make this determination, is itself of the same sort. And this insight was capable, from that point onward, of freeing me from the repentence and remorse that commonly agitate the consciences of these frail and irresolute minds that allow themselves to go about with inconstancy, treating things as if they were good, only to judge them later to be bad.

My third maxim was always to try to conquer myself rather than fortune, to change my desires rather than the order of the world; and generally to become accustomed to believing that there is nothing that is utterly within our power, except for our thoughts, so that, after having done our best regarding things external to us, everything that fails to bring us success, from our point of view, is absolutely impossible. And this principle alone seemed sufficient to stop me from desiring anything in the future that I would not acquire, and thus seemed sufficient to make me contented. For, *26* our will tending naturally to desire only what our intellect represents to it as in some way possible, it is certain that, if we consider all of the goods that are outside us as equally beyond our power, we should have no more regrets about lacking what seems owed to us at birth, when we are deprived of them through no fault of our own, than we should have for not possessing the kingdoms of China or Mexico. Thus, making a virtue of necessity, as they say, we shall no more desire to be healthy if we are sick, or to be free if we are in prison, than we would desire to have a body made of matter as incorruptible as diamonds, or wings with which to fly like birds. But I confess that long exercise is needed as well as frequently repeated meditation in order to become accustomed to looking at everything from this point of view; and I believe that in this principally lay the secret of the philosophers who at one time were able to free themselves from fortune's domination and who could, despite their sorrows and their poverty, rival their gods in their happiness. For occupying themselves ceaselessly with considering the limits prescribed to them by nature, they so perfectly persuaded themselves that nothing was in their power save their own thoughts, that this alone was sufficient to stop them from having feelings about any other objects; and they controlled their thoughts so absolutely, that they thereby had some reason for judging themselves richer, more powerful, freer, and happier than those other men who, not having this philosophy—however favored by nature and fortune as they *27* may be—never controlled everything they wished to control.

Finally, to conclude this code of morals, I thought it advisable to review the various occupations that men take up in this life, so as to try to choose the best one; and, not wanting to say anything about the occupations of others, I believed I could not do better than to continue in the occupation I was in at that time, namely cultivating my reason all my life and advancing, as best as I could, in the knowledge of truth, following the method I had prescribed to myself. I had met with such intense satisfaction, since the time I had begun to make use of this method, that I did not believe one could receive sweeter or more innocent satisfaction in this life; and, discovering every day by its means some truths that seemed important to me and commonly ignored by other men, I had a satisfaction that so filled my mind that nothing else was of any consequence to me. In addition, the three preceding maxims were founded merely on the plan I had of continuing my self-instruction; for since God has given each of us a certain light by which to distinguish the true from the false, I should not believe I ought to be content for a single moment with the opinions of others, had I not proposed to use my own judgment to examine them when there was time; and I should not have been able to be free of scruple in following these opinions, had I not hoped I would not waste the opportunity thereby *28* of finding better ones, in case there were better ones. And finally, I could not have curbed my desires or have been contented, had I not followed a road by which, believing I was assured of acquiring all the knowledge of which I was capable, I believed I was assured of acquiring by the same means all the true goods that would ever be in my power; given that our will neither pursues nor flees an object unless our intellect represents that object to the will as either good or bad, it suffices to judge well in order to do well, and to judge as best one can, in order also to do one's best, that is, to acquire all the virtues and, along with them, all the other goods that one can acquire; and while one is certain that this is the case, one could not fail to be contented.

After having assured myself of these maxims and having put them aside, along with the truths of the faith, which have always held first place in my set of beliefs, I judged that, as far as the rest of my opinions were concerned, I could freely undertake to rid myself of them. And insofar as I hoped I could be more successful in social interchange than in remaining any longer shut up in the small stove-heated room where I had had all of these thoughts, I set out again on my travels, the winter not yet completely over. And in all the following nine years I did nothing but wander here and there about the world, trying to be more a spectator than an actor in all the comedies that were being played out there; and reflecting particularly in each matter on what might render it suspect and give us occasion for error, I meanwhile rooted out from my mind all the errors that had been able to creep in undetected. Not that I was thereby aping the sceptics *29* who doubt merely for the sake of doubting and put on the affectation of perpetual indecision; for, on the contrary, my entire plan tended simply to give me assurance and to reject shifting ground and sand so as to find rock or clay. In this I was quite successful, it seems to me, inasmuch as, trying to discover the falsity or uncertainty of the propositions I was examining—not by feeble conjectures but by clear and certain reasonings—I never found anything that was so doubtful that I could not draw some rather certain conclusion from it, even if it were merely that it contained nothing certain. And just as in tearing down an old house, one usually saves the wreckage for use in building the new house, similarly, in destroying all of those opinions that I judged to be poorly supported, I made various observations and acquired many experiences that I later found useful in establishing opinions that were more certain. Moreover, I continued to practice the method I had prescribed for myself; for besides taking care generally to conduct all my thoughts according to the rules of this method, from time to time I set aside a few hours that I spent in practicing the method on mathematical problems, or even in various other problems that I could render similar to those of mathematics, by detaching them from all the principles of the other sciences, which I did not find to be sufficiently firm, as you will observe I have done in several cases that are explained in this volume. And thus, without living any other way in outward appearance than those who, having no other task but living sweet *30* and innocent lives, are eager to separate pleasures from vices and who, to enjoy their leisure without becoming bored, engage in all sorts of honorable diversions, I did not cease to pursue my plan and to profit in the knowledge of the truth, perhaps more than if I had done nothing but read books or keep company with men of letters.

All the same, these nine years slipped away before I had as yet taken any stand regarding the difficulties commonly debated by learned men, or had begun to seek the foundations of any philosophy that was more certain than the commonly accepted one. And the example of many excellent minds, which, having already had this plan, appeared to me not to have succeeded, made me conjure thoughts of so many difficulties that perhaps I should not yet have dared to try it if I had not seen that some people had already passed the rumor around that I had already succeeded. I cannot say on what they based this opinion; and if I have contributed anything to this by my conversations, it must have been more because I admitted what

I did not know more ingenuously than do those who have studied only a little, and perhaps also because I showed the reasons I had for doubting many of the things that other people regard as certain, than because I was boasting of any knowledge. But being decent enough not to want someone to take me for something other than I was, I thought it necessary to try by *31* every means to make myself worthy of the reputation bestowed upon me; and it is exactly eight years since this desire made me resolve to take my leave of all those places where I could have acquaintances, and to retire here, in a country where the long duration of the war has established such well-ordered discipline that the armies quartered there seem to be there solely for the purpose of guaranteeing the enjoyment of the fruits of peace with even greater security, and where among the crowds of a great and very busy people and more concerned with their own affairs than curious about the affairs of others, I have been able to live as solitary and as retired a life as I could in the remotest deserts—but without lacking any of the amenities that are to be found in the most populous cities.

Part Four

I do not know whether I ought to tell you about the first meditations I made there; for they are so metaphysical and so out of the ordinary, that perhaps they would not be to everyone's liking. Nevertheless, so that one might be able to judge whether the foundations I have laid are sufficiently firm, I am in some sense forced to speak. For a long time I have noticed that in moral matters one must sometimes follow opinions that one knows are quite uncertain, just as if they were indubitable, as has been said above; but since then I desired to attend only to the search for truth, I thought it necessary that I do exactly the opposite, and that I reject as absolutely false everything in which I could imagine the least doubt, so as to see whether, after this process, anything in my set of beliefs remains that is *32* entirely indubitable. Thus, since our senses sometimes deceive us, I decided to suppose that nothing was exactly as our senses would have us imagine. And since there are men who err in reasoning, even in the simplest matters in geometry, and commit paralogisms, judging that I was just as prone to err as the next man, I rejected as false all the reasonings that I had previously taken for demonstrations. And finally, taking into account the fact that the same thoughts we have when we are awake can also come to us when we are asleep, without any of the latter thoughts being true, I resolved to pretend that everything that had ever entered my mind was no more true than the illusions of my dreams. But immediately afterward I noticed that, during the time I wanted thus to think that everything was false, it was necessary that I, who thought thus, be something. And noticing that this truth—*I think, therefore I am*—was so firm and so certain that the most extravagant suppositions of the sceptics were unable to shake it, I judged that I could accept it without scruple as the first principle of the philosophy I was seeking.

Then, examining with attention what I was, I saw that I could pretend that I had no body and that there was no world nor any place where I was, but that I could not pretend, on that account, that I did not exist; and that, on the contrary, from the very fact that I thought about doubting the truth of other things, it followed very evidently and very certainly that I existed. On the other hand, had I simply stopped thinking, even if all the rest of *33* what I have ever imagined were true, I would have no reason to believe that I existed. From this I knew that I was a substance the whole essence or nature of which was merely to think, and which, in order to exist, needed no place and depended on no material thing. Thus this "I," that is, the soul through which I am what I am, is entirely distinct from the body, and is even easier to know than the body, and even if there were no body, the soul would not cease to be all that it is.

After this, I considered in a general way what is needed for a proposition to be true and certain; for since I had just found a proposition that I knew was true, I thought I ought also know in what this certitude consists. And having noticed that there is nothing in all of this—*I think, therefore I am*—that assures me that I am uttering the truth, except that I see very clearly that, in order to think, one must exist; I judged that I could take as a general rule that the things we conceive very clearly and very distinctly are all true, but that there only remains some difficulty in properly discerning which are the ones that we distinctly conceive.

Following this, reflecting upon the fact that I doubted and that, as a consequence, my being was not utterly perfect (for I saw clearly that it is a greater perfection to know than to doubt), I decided to search for the source from which I had learned to think of a thing more perfect than myself; and I readily knew that this ought to originate from some nature *34* that was in effect more perfect. As to those thoughts of mine that were of many other things outside me—such as the sky, the earth, light, heat, and a thousand other things—I was not quite so anxious to know where they came from, since, having noticed nothing in them that seemed to me to make them superior to me, I could believe that, if they were true, they were dependencies of my nature, to the extent that it had any perfection; and that if they were not true, I received them from nothing, that is, they were in me because I had some defect. But the same could not hold for the idea of a being more perfect than my own; for the receiving of this idea from nothing is a manifest impossibility; and since it is no less a contradiction that something more perfect should follow from and depend upon something less perfect than that something can come from nothing, I certainly could not obtain it from myself. It thus remained that this idea was placed in me by a nature truly more perfect than I was, and even that it had within itself all the perfections of which I could have any idea, that is, to put my case in a single word, that this nature was God. To this I added that, since I knew of some perfections that I did not possess, I was not the only being in existence (here, if you please, I shall use freely the language of the School), but that of necessity it must be the case that there is something else more perfect, upon which I depended, and from which I acquired all that I had. For, had I been alone and independent of *35* everything else, so as to have derived from myself all of that small allotment of perfection I had through participation in the perfect being, I would have been able for the same reason to give myself the remainder of what I knew was lacking in me; and thus I would be infinite, eternal, unchanging, all-knowing, all-powerful—in short, I would have all the perfections I could discern in God. For, following from the reasonings I have just given, to know the nature of God, as far as my own nature was able, I had only to consider each thing about which I found an idea in myself, whether or not it was a perfection to have them, and I was certain that none of those that were marked by any imperfection were in this nature, but that all other perfections were. So I observed that doubt, inconstancy, sadness and the like could not be in him, given the fact that I would have been happy to be exempt from them. Now, over and above that, I had ideas of several sensible and corporeal things; for even supposing that I was dreaming and that everything I saw or imagined was false, I still could not deny that the ideas were not truly in my thought. But since I had already recognized very clearly in my case that intelligent nature is distinct from corporeal nature, taking into consideration that all composition attests to dependence and that dependence is manifestly a defect, I therefore judged that being composed of these two natures cannot be a perfection in God and that, as a consequence, God is not thus composed. But, if there are bodies in *36* the world, or intelligences, or other natures that were not entirely perfect, their being ought to depend on God's power, inasmuch as they cannot subsist without God for a single moment.

After this, I wanted to search for other truths, and, having set before myself the object dealt with by geometricians, which I conceived to be like a continuous body or a space indefinitely extended in length, breadth, and height or depth, divisible into various parts which could have various shapes and sizes and be moved or transposed in all sorts of ways (for all this the geometricians take for granted in their object), I ran through some of their simplest proofs. And having noticed that this great certitude that everyone attributes to them is founded only on the fact that one conceives them evidently conforming to the rule that I mentioned earlier, I also noted that there had been nothing in them that assured me of the existence of their object. For I saw very well that by supposing, for example, a triangle, it is necessary for its three angles to be equal to two right angles; but I did not see anything in all this which would assure me that any triangle existed. On the other hand, returning to an examination of the idea I had of a perfect being, I found that existence was contained in it, in the same way as the fact that its three angles are equal to two right angles is contained in the idea of a triangle, or that, in the case of a sphere, all its parts are equidistant from its center, or even more evidently so; and consequently, it is, at the very least, just as certain that God, who is a perfect being, is or exists, as any demonstration in geometry could be.

But what makes many people become persuaded that it is difficult to *37* know this (i.e., the existence of the perfect being), and also even to know what kind of thing their soul is, is that they never lift their minds above sensible things and that they are so much in the habit of thinking about only what they can imagine (which is a particular way of thinking appropriate only for material things), that whatever is not imaginable seems to them to be unintelligible. This is obvious enough from what even the philosophers in the Schools take as a maxim: that there is nothing in the understanding that has not first been in the senses (where obviously the ideas of God and the soul have never been). And it seems to me that those who want to use their imagination to comprehend these things are doing the same as if, to hear sounds or to smell odors, they wanted to use their eyes, except for this difference: the sense of sight assures us no less of the truth of its objects than do the senses of smell or hearing, whereas neither our imagination nor our sense could ever assure us of anything if our understanding did not intervene.

Finally, if there are men who have not yet been sufficiently persuaded of the existence of God and their soul by means of the reasons I have brought forward, I would very much like them to know that all the other things they thought perhaps to be more certain—such as having a body, there being stars and an earth, and the like—are less certain. For although one might have a moral certainty about these things, which is such that it

38 seems outrageous for anyone to doubt it, yet, while it is a question of metaphysical certitude, it seems unreasonable for anyone to deny that there is a sufficient basis for one's not being completely certain about the subject, given that one can, in the same fashion, imagine that while asleep one has a different body and that one sees different stars and a different earth, without any of it being the case. For how does one know that the thoughts that come to us in our dreams are more false than the others, given that often they are no less vivid or express? Let the best minds study this as much as they please, I do not believe they can give any reason that would suffice to remove this doubt, were they not to presuppose the existence of God. For first of all, even what I have already taken for a rule—namely that all the things we very clearly and very distinctly conceive are true—is certain only because God is or exists, and is a perfect being, and because all that is in us comes from him. Thus it follows that our ideas or our notions, being real things and coming from God, insofar as they are clear and distinct, cannot to this extent fail to be true. Thus, if we have ideas sufficiently often that contain falsity, this can only be the case with respect to things that have something confused or obscure about them, since in this regard they participate in nothing; that is, they are thus in us in such a confusion only because we are not perfect. And it is evident that there is no less a contradiction that falsity or imperfection, as such, *39* proceed from God, than that truth or perfection proceed from nothing. But if we did not know that all that is real and true in us comes from a perfect and infinite being, however clear and distinct our ideas may be, we would have no reason that assured us that they had the perfection of being true.

But after the knowledge of God and the soul has thus rendered us certain of this rule, it is very easy to know that the dreams we imagine while asleep ought in no way make us doubt the truth of the thoughts we have while awake. For if it should happen, even while one is asleep, that someone has a very distinct idea, as, for example, when a geometrician invents a new demonstration, his being asleep does not impede its being true. And as to the most common error of our dreams, which consists in the fact that they represent to us various objects in the same way as our exterior senses do, it is of no importance that it gives us the occasion to question the truth of such ideas, since they can also deceive us just as often without our being asleep—as when those with jaundice see everything as yellow-colored, or when the stars or other distant bodies appear to us a great deal smaller than they are. In short, whether awake or asleep, we should never allow ourselves to be persuaded except by the evidence of our reason. And it is to be noted that I said this of our reason, and not of our imagination or our senses. For, although we see the sun very clearly, *40* we should not on that account judge that it is only as large as we see it; and we can very well imagine distinctly the head of a lion grafted on the body of a goat, without necessarily concluding for that reason that there existed a chimera; for reason does not suggest to us that what we thus see or imagine is true. But it does suggest to us that all our ideas or notions ought to have some foundation in truth; for it would not be possible that God, who is all perfect and entirely truthful, would have put them in us without that. And because our reasonings are never so evident nor so complete while we are asleep as they are while we are awake, even though our imaginations are sometimes just as, or even more, vivid and express when asleep, reason also suggests to us that our thoughts are unable all to be true, since we are not all-perfect; what truth there is in them ought infallibly to be found in those we have when awake rather than those we have in our dreams.

Part Five

I would be quite happy to continue and to show here the whole chain of other truths that I had deduced from these first ones. But since, to that end, it would now be necessary for me to speak about many problems that are a matter of controversy among the learned, with whom I do not want to get into a scuffle, I believe it would be better for me to abstain and to state only in a general way what they are, so that it might be left to the most wise to judge whether it be useful for the public to be more informed as to the particulars. I have always remained firm in my resolve not to *41* suppose any principle but the one I just used to demonstrate the existence of God and the soul, and to take nothing to be true that does not seem to me clearer and more certain than have the demonstrations of the geometricians been previously. And still I dare say not only that I have found the means of satisfying myself in a short time regarding all the main difficulties commonly treated in philosophy, but also that I have noted certain laws that God has so established in nature and has impressed in our souls such notions of these laws that, after having reflected sufficiently, we cannot deny that they are strictly adhered to in everything that exists or occurs in the world. Now, in considering the chain of these laws, it seems that I have discovered several truths more useful and more important than all I had previously learned or even hoped to learn.

But because I have tried to explain these principles in a treatise that certain considerations kept me from publishing,[5] I could not make them better known than by declaring here in summary form what the treatise contains. I had intended to include in it everything I thought I knew, before writing it down, concerning the nature of material things. But just as painters who, unable to represent equally on a flat surface all the various sides of a solid body, choose one of the principal sides which they place *42* alone in the light of day, and, darkening with shadows all the rest, make them appear only as they can be seen while someone is looking at the principal side; just so, fearing I could not put in my discourse all that I had thought on the matter, I tried simply to speak at length about what I conceived with respect to light; then, at the proper time, to add something about the sun and the fixed stars, since light originates almost entirely from them; something about the heavens since they transmit light; about the planets, comets, and the earth since they reflect light; and particularly, about all terrestrial bodies, since they are either colored, transparent, or luminous; and finally, about man, since he is the observer of all this. But, to put all these things in a slightly softer light and to be able to say more freely what I have judged in these matters, without being obliged either to follow or to refute the opinions accepted among the learned, I here resolved to leave all this world to their disputes and to speak only of what would happen in a new world, were God now to create enough matter to make it up, somewhere in imaginary space, and if he were to put in motion variously and without order the different parts of this matter, so that he concocted as confused a chaos as the poets could ever imagine and that later he did no more than apply his ordinary conserving activity to nature, letting nature act in accordance with the laws he has established. Thus, first, I described this matter and tried to represent it such that there is nothing in the world, it seems to me, more clear and more intelligible, *43* with the exception of what has already been said about God and the soul; for I even supposed explicitly that there was in it none of those forms or qualities about which disputes occur in the Schools, nor generally anything the knowledge of which was not so natural to our souls that one cannot even pretend to ignore it. Moreover, I showed what were the laws of nature; and without supporting my reasons on any other principle but the infinite perfections of God, I tried to demonstrate all the laws about which

5. Descartes's *Le Monde (The World).* See René Descartes, *Le Monde ou Traité de la lumière,* translation and introduction by Michael Sean Mahoney (New York: Abaris Books, Inc., 1979). One of the considerations preventing the publication of *Le Monde* was the trial in 1633 of Galileo by the Holy Office in Rome.

one might have been able to doubt and to show that they were such that, even if God had created several worlds, there could have been none in which these laws failed to be observed. Next, I showed how most of the matter of this chaos must, in observance of these laws, dispose and arrange itself in a certain way that makes it similar to our heavens; how, meanwhile, some of its parts should form an earth; others, planets and comets; and still others a sun and fixed stars. And here, dwelling on the subject of light, I explained at great length what this light was that ought to be found in the sun and the stars, and how thence it coursed in an instant the immense stretches of celestial space, and how it was reflected from the planets and comets to the earth. To that I added also several things touching on the substance, position, movements, and all the various qualities of these heavens and stars; and as a result I thought that I said enough on the matter to show that there is nothing to be mentioned in this world which should not, or at least could not, appear utterly similar to the world *44* I have just described. Next I went on to speak particularly of the earth: how, although I had expressly supposed that God had not put any weight in the matter out of which the earth was composed, still none of its parts would cease to tend precisely toward its center; how, having water and air on its surface, the disposition of the heavens, the stars, and principally the moon ought to cause an ebb and flow that would be similar in all respects to what we observe in our own seas, and, what is more, a certain coursing—as much of the water as of the air—from east to west, such as is observed between the tropics; how mountains, seas, springs, and rivers could be formed naturally therein and how metals could have made their way naturally into mines; how plants could have grown naturally in the fields and generally how all the bodies one calls mixed or composed could have been engendered naturally. And, among other things, since over and above the stars I know of nothing else in the world that produces light except fire, I tried to make clearly understood all that belonged to its nature, how it occurs, how it feeds, how sometimes there is only heat but no light, and sometimes only light but no heat; how it can introduce various colors into various bodies, and the same for various other qualities; how it melts some things and hardens others; how it can consume nearly everything or turn it into ashes and smoke; and finally, how from these ashes, merely by the violence of its action, it produces glass; for since this transmutation of ashes into glass seems to me to be as awesome as any other event in nature, *45* I took particular pleasure in describing it.

Yet I did not want to infer from all these things that this world has been created in the manner I described, for it is much more likely that, from the very beginning, God made it such as it was supposed to be. But it is certain (and this is an opinion commonly held among theologians) that the action by which God conserves the world is precisely the same action by which he created it; so that even if he had never given it, at the beginning, any other form but that of chaos, provided he established the laws of nature and applied his conserving activity to make nature function just as it does ordinarily, one can believe—without belittling the miracle of creation—that by such activity alone all the things that are purely material could have been able, as time went on, to make themselves just as we now see them. And their nature is much easier to conceive, when one sees them gradually coming to be in this manner, than when one considers them only in their completed state.

From the description of inanimate bodies and plants I passed to the description of animals and particularly of human beings. But since I had insufficient knowledge in this area to speak in the same way as I did in the rest, that is in demonstrating effects by their causes and showing from what seeds and in what way nature should produce them, I contented myself with supposing that God formed the body of a man entirely similar *46* to one of ours, as much in the outward shape of its members as in the internal arrangement of its organs, without making it out of any material other than the type I had described, and without putting in it, at the start, any rational soul, or anything else to serve as a vegetative or sensitive soul, but merely exciting in the man's heart one of those fires without light that I had already explained, and which, having no other nature but that which heats up hay when it has been bundled up before drying, or which boils new wines while they are left to ferment on the stalk. For examining the functions that could, consequently, be in this body, I found precisely all the things that could be in us without our thinking about them, and hence without our soul (that is, that part distinct from the body of which I have said before that its nature is only to think) contributing anything to them, and these are all the same so that one can say that nonrational animals resemble us. But I could not on that account find any of those things that, being dependent on thought, are the only things that belong to us insofar as we are men, although I found them all later when I had supposed that God created a rational soul and that he joined it to this body in the particular way I have described.

But so that one might see the way in which I treated this matter, I want to place here the explanation of the movement of the heart and arteries; because this is the first and most general movement that one observes in *47* animals, one easily judges on the basis of it what one ought to think regarding all the rest. So that one might have less difficulty in understanding what I shall say on the matter, I would like those who are not versed in anatomy to take the trouble, before reading this, to have dissected in their presence the heart of a large animal that has lungs (for it is in many respects sufficiently similar to that of a man), and to be shown the two chambers or ventricles in the heart. First, the one on the right side of the heart, into which two very large tubes lead, namely the *vena cava,* which is the main receptable for blood, somewhat like the trunk of a tree all of whose other veins are branches, and the *vena arteriosa* (which has been rather ill-named, since it is, after all, an artery), which, taking its origin from the heart, breaks up, on leaving the heart, into several branches that are spread all throughout the lungs. Now the chamber on the left side, into which two tubes lead in the very same fashion, which are just as large or larger than the other tubes: namely, the *arteria venosa* (which also has been ill-named, since it is only a vein), which comes from the lungs where it is divided into many branches interlaced with those of the *vena arteriosa* and with those in the passageway called the windpipe, through which enters the air one breathes; and the aorta which, on leaving the heart, sends its branches all over the body. I would also like to have one be carefully shown the eleven little valves that, like so many little doors, open and shut the four openings in the two ventricles: namely, three at the entrance to the *vena cava,* where they are so disposed that they cannot stop *48* the blood it contains from flowing into the right ventricle of the heart, and yet stop any of it from being able to leave the ventricle; three at the entrance to the *vena arteriosa* which, being disposed in just the opposite way, readily allow the blood in this ventricle to pass into the lungs, but do not allow any blood in the lungs to return to this ventricle; two others at the entrance to the *arteria venosa,* which let blood flow from the lungs to the left ventricle of the heart but resist its returning; and three at the entrance to the aorta that permit blood to leave the heart but stop it from returning. And there is no need to search for any other reason for the number of valves except that the opening of the *arteria venosa,* being oval-shaped because of its location, can be conveniently closed with two, while the others, being round, can better be closed with three. Moreover, I would like people to consider that the aorta and the *vena arteriosa* are of a much sturdier and firmer constitution than the *arteria venosa* and the *vena cava;* and that these last two are enlarged before entering the heart and form, as it were, sacks, called the auricles of the heart, which are made of flesh similar to that of the heart; and that there is always more heat in the heart than anywhere else in the body; and, finally, that this

heat is able to bring it about that, if a drop of blood should enter its
49 ventricles, the blood expands forthwith and is dilated, just as all liquids generally do, when one lets them fall drop by drop in a very hot vessel.

For, after that, I have no need of saying anything else to explain the movement of the heart, except that, while its ventricles are not full of blood, blood necessarily flows from the *vena cava* into the right ventricle and from the *arteria venosa* into the left ventricle—given that these two vessels are always full, and their apertures, which open toward the heart, cannot then be shut. But as soon as two drops of blood have thus entered the heart, one into each ventricle, these drops—which can only be very large, because the openings through which they enter are very large and the vessels whence they come are quite full of blood—are rarefied and dilated, because of the heat they find there, by means of which, making the whole heart inflate, they push and close the five little doors that are at the entrances of the two vessels whence they come, impeding any further flow of blood into the heart; and continuing to be more and more rarified, they push and open the six other little doors that are at the entrances of the two other vessels by which they leave; by this means they inflate all the branches of the *vena arteriosa* and the aorta, almost at the same instant as the heart, which immediately afterward contracts—as do the arteries too—since the blood that has entered there gets cooled again and their six small doors close, and the five little doors of the *vena cava* and the
50 *arteria venosa* are opened up again and grant passage to two other drops of blood which immediately inflate again the heart and the arteries, the same as before. And because the blood that thus enters the heart passes by the two sacks called auricles, so it follows that their movement is contrary to the heart's: they are deflated while the heart is inflated. For the rest (so that those who do not know the force of mathematical demonstrations and are not in the habit of distinguishing true reasons from apparent reasons should not venture to deny this without examining it), I want to put them on notice that this movement that I have been explaining follows just as necessarily from the mere disposition of the organs that can be seen in the heart by the unaided eye and from the heat that can be felt with the fingers, and from the nature of blood which can be known through experience, as do the motions of a clock from the force, placement, and shape of its counterweights and its wheels.

But if one asks how it is that blood in the veins is not exhausted, in flowing continually into the heart, and how it is that the arteries are not overly full, since all the blood that passes through the heart goes there, I need only answer with what has already been written by an English physician,[6] to whom must be given homage for having broken the ice in this area, and who was the first to have taught that there are many small passages at the extremities of the arteries through which the blood they receive enters in the small branches of the veins, from which it heads once more for the heart, so that its course is merely a perpetual circulation. He proves this very effectively from the common experience of surgeons, who, *51* on binding an arm moderately tightly above the spot where they opened the vein, cause blood to flow out in even greater abundance than if they had not bound the arm. And the opposite would happen if they bound the arm below, between the hand and the opening, or if they tied it above very tightly, because it is clear that a tourniquet tied moderately, being able to stop blood already in the arm from returning to the heart through the veins, does not on that account stop any new blood that is always coming in from the arteries, since they are located below the veins, and their valves, being tougher, are less easy to press, and since the blood coming from the heart tends to pass with greater force through the arteries toward the hand than it does on returning to the heart through the veins. Because the blood flows from the arm through the opening in one of the veins, there must necessarily be some passages below the tourniquet, that is, toward the extremities of the arm, through which it can come from the arteries. He also proves quite effectively what he says regarding the flow of blood, first, through certain small valves that are so disposed in various places along the length of the veins that they do not allow blood to pass from the middle of the body to the extremities, but only to return from the extremities toward the heart; and, second, through experience which shows that all the blood in the body can flow out of it in a very short amount of time through just one artery if it is cut open, even if it is very tightly bound quite close to the heart, and cut open between the heart and the tourniquet, so that one could not have any basis for imagining that the *52* blood left from anywhere but the heart.

But there are many other things that attest to the fact that the true cause of this movement of blood is what I said it is. First, the difference that one notices between blood leaving the veins and blood leaving the arteries can arise only from the fact that it is rarified and, so to speak, distilled, in passing through the heart; it is more subtle, more lively, and warmer just after having come out of the heart, that is, while it is in the arteries, than

6. William Harvey (1578–1657), English physiologist who demonstrated the function of the heart as a kind of pump and the complete circulation of blood throughout the body. His most important work is *Anatomical Exercises on the Motion of the Heart and Blood* (1628).

it is shortly before it enters the heart, that is, while it is in the veins. If one takes a good look, one will find that this difference appears more clearly near the heart and not so much in those places furthest removed from the heart. Now the toughness of these valves, which compose the *vena arteriosa* and the aorta, shows quite well that the blood bangs against them with more force than it does against the veins. And why are the left ventricle of the heart and the aorta more spacious and larger than the right ventricle and the *vena arteriosa,* unless it is that the blood in the *arteria venosa,* having been only in the lungs after having passed through the heart, is more subtle and is more forcefully and easily rarified than what comes immediately from the *vena cava?* And what can physicians divine from taking the pulse, if they do not know that, as the blood changes its nature, it can be rarified by the heat of the heart with more or less strength, with more *53* or less liveliness than before? And if one examines how this heat is communicated to the other members, must one not admit that it is by means of the blood which, on passing through the heart, is warmed and from that point is spread throughout the whole body? It follows from this that if one removes the blood from some part, one also removes the heat; and even if the heart were as intensely hot as a piece of glowing iron, it would not be enough to heat the feet and hands as it does, unless it continuously sent new supplies of blood. Now one also knows from this that the true purpose of respiration is to bring sufficient quantities of fresh air into the lungs, to cause the blood, which leaves the right ventricle of the heart where it was rarified and, as it were, changed into vapors, to be condensed and to be converted once again into blood, before returning to the left ventricle; without this process the blood could not properly aid in nourishing the fire that is in the heart. This is confirmed in the fact that one sees that animals without lungs have but one single ventricle in their hearts, and that children, who cannot use their lungs while locked within their mother's womb, have an opening through which blood flows from the *vena cava* into the left ventricle of the heart, as well as a tube through which the blood goes from the *vena arteriosa* to the aorta, without passing through the lungs. Now how does digestion take place in the stomach if the heart does not send heat there through the arteries, along with some of the more fluid parts of the blood that help dissolve the food placed there? And is it not easy to understand the action that changes the juices of the food into blood, if one considers that they are distilled, in passing and repassing through the heart, perhaps more than one or two hundred *54* times a day? And need anything else be said to explain nutrition and the production of the body's various humors, except that the force with which the blood, as it is being rarified, passes from the heart to the extremities of the arteries, makes some of its parts stop in those of the members where they are found and there take the place of others that they expel; and that, according to their location or shape, or the smallness of the pores they encounter, they tend to go some places rather than others, in just the same way, as anyone can see, as various sieves that, being variously perforated, help separate out different size grains from one another? And finally, what is most remarkable in all this is the generation of animal spirits that are like a very subtle wind, or better, like a very pure and lively flame that rises continually in great abundance from the heart to the brain, and from there goes through the nerves into the muscles, and gives movement to all the members, without the need for imagining any other reason for the fact that the parts of blood which, being the most agitated and most penetrating, are the most likely to make up these spirits, go to the brain rather than elsewhere, except that the arteries that carry these parts of blood are those that go from the heart in the straightest line of all; for, according to the laws of mechanics, which are the same as the laws of nature, when several objects tend all together to move in a single direction, where there is not enough room for all of them, as the parts of blood leaving the left ventricle of the heart tend toward the brain, the weakest and least lively must be pushed aside by the stronger which in this way *55* arrive by themselves at the brain.

I provided a sufficiently detailed explanation for all these things in the treatise that I had intended earlier to publish. And I went on to show of what sort the fabric of the nerves and muscles of the human body must be, so that the animal spirits within might have the force to move its members; thus one observes that heads, shortly after being severed, still move about, and bite the earth, even though they are no longer alive. I also showed what changes ought to take place in the brain to cause wakefulness, sleep, and dreams; how light, sounds, odors, tastes, heat, and all the other qualities of external objects can imprint various ideas through the medium of the senses; how hunger, thirst, and the other internal passions can also send their own ideas; what needs to be taken for the common sense, where these ideas are received, for the memory which conserves them, and for the imagination which can change them in various ways and make new ones out of them, and by the same means, distributing the animal spirits in the muscles, make the members of this body move in as many different ways, each appropriate to the objects presented to the senses and to the internal passions that are in the body, as our own bodies can move, without being led to do so through the

intervention of our will. This ought not seem strange to those who, cognizant of how many different automata or moving machines the ingenu-
56 ity of men can devise, using only a very small number of parts, in comparison to the great multitude of bones, muscles, nerves, arteries, veins, and all the other parts which are in the body of each animal, will consider this body like a machine that, having been made by the hand of God, is incomparably better ordered and has within itself movements far more admirable than any of those machines that can be invented by men.

And I paused here particularly to show that, if there were such machines having the organs and the shape of a monkey or of some other nonrational animal, we would have no way of telling whether or not they were of the same nature as these animals; if instead they resembled our bodies and imitated so many of our actions as far as this is morally possible, we would always have two very certain means of telling that they were not, for all that, true men. The first means is that they would never use words or other signs, putting them together as we do in order to tell our thoughts to others. For one can well conceive of a machine being so made as to pour forth words, and even words appropriate to the corporeal actions that cause a change in its organs—as, when one touches it in a certain place, it asks what one wants to say to it, or it cries out that it has been injured, and the like—but it could never arrange its words differently so
57 as to answer to the sense of all that is said in its presence, which is something even the most backward men can do. The second means is that, although they perform many tasks very well or perhaps can do them better than any of us, they inevitably fail in other tasks; by this means one would discover that they do not act through knowledge, but only through the disposition of their organs. For while reason is a universal instrument that can be of help in all sorts of circumstances, these organs require a particular disposition for each particular action; consequently, it is morally impossible for there to be enough different devices in a machine to make it act in all of life's situations in the same way as our reason makes us act.

For by these two means one can know the difference between men and beasts. For it is very remarkable that there are no men so backward and so stupid, excluding not even fools, who are unable to arrange various words and to put together discourse through which they make their thoughts understood; but, on the other hand, there is no other animal, perfect and well bred as it may be, that can do likewise. This is not due to the fact that they lack the organs for it, for magpies and parrots can utter words just as we can, and still they cannot speak as we can, that is, by giving evidence of the fact that they are thinking about what they are saying; although the deaf and dumb are deprived just as much as—or more than—animals of the organs which aid others in speaking, they are 58 in the habit of inventing for themselves various signs through which they make themselves understood to those who are usually with them and have the leisure to learn their language. And this attests not merely to the fact that animals have less reason than men but that they have none at all. For one sees that not much of it is needed so as to be able to speak; given that one notices an inequality among animals of the same species, just as is the case among men, and that some are easier to train than others, one could not believe that a monkey or a parrot which is the most perfect of its species would not equal in this one of the most stupid of children, or at least a child with a disturbed brain, unless their soul were of an utterly different nature from our own. And one should not confuse words with natural movements that display the passions and can be imitated by machines as well as by animals; nor should we think, as did some ancients, that animals speak, although we do not understand their language; for if that were true, since they have many organs similar to our own, they could also make themselves understood by us just as they are by their peers. It is also remarkable that, although there are many animals that show more inventiveness than we do in some of their actions, one nevertheless sees that they show none at all in many other actions; consequently, the fact that they do something better than we do does not prove that they have a mind; for were this the case, they would be more rational than any of us and would excel us in everything; but rather it proves that they do not 59 have a mind, and that it is nature that acts in them, according to the disposition of their organs—just as one sees that a clock made only of wheels and springs can count the hours and measure time more accurately than we can with all our powers of reflective deliberation.

After this, I described the rational soul and showed that it can in no way be drawn from the potentiality of matter, as can the other things I have spoken of, but that it ought expressly to be created; and how it is not enough for it to be lodged in the human body, like a pilot in his ship, unless perhaps to move its members, but it must be joined and united more closely to the body so as to have, in addition, feelings and appetites similar to our own, and thus to make up a true man. As to the rest, I elaborated here a little about the subject of the soul because it is of the greatest importance; for, after the error of those who deny the existence of God (which I believe I have sufficiently refuted), there is nothing that puts weak minds at a greater distance from the straight road of virtue than imagining that the soul of animals is of the same nature as ours and that, as a consequence, we have no more to fear nor to hope for after this life than have flies or ants; whereas, while one understands how much they differ, one understands much better the reasons that prove that our soul is of a nature entirely independent of the body and, consequently, that it *60* is not subject to die with it. Now, since one cannot see any other causes that might destroy it, one is naturally led to judge from this that the soul is immortal.

Part Six

But it is now three years since I completed the treatise containing all these things, and I began to review it in order to put it into the hands of a publisher, when I learned that people to whom I defer, and whose authority over my actions cannot be less than that of my reason over my thoughts, had disapproved of a certain opinion in the realm of physics, published a short time before by someone else,[7] concerning which I do not wish to say that I was in agreement, but rather that I had found nothing in it, before their censuring of it, that I could imagine to be prejudicial either to religion or to the state, nor had I found anything that would have stopped me from writing it, had reason persuaded me of it; and this made me fear that I might likewise find among my opinions some in which I was mistaken, notwithstanding the great care that I had always taken not to accept into my set of beliefs any new opinions for which I did not have very certain demonstrations and never to write anything that could turn to someone's disadvantage. This was sufficient to make me change the resolution I had made to publish them. For although the reasons for which I had earlier made the resolution were very strong, my inclination, always making me hate the business of writing books, immediately made me find enough *61* other reasons to excuse me from it. And these reasons, both for and against, were such that not only do I have an interest in saying these things, but perhaps the public too has an interest in knowing them.

I had never made much of what came from my mind, and as long as I had reaped no other fruits from the method which I used, aside from my

7. Galileo Galilei (1564–1642), Italian astronomer, mathematician and physicist. His *Dialogue . . . on the Two Chief Systems of the World* (1632), in which he advanced the theory of the movement of the earth, occasioned the Inquisitors of the Holy Office to conduct a trial in Rome and to extort a retraction of that theory from Galileo. Descartes, who also advocated a theory of terrestrial motion, was not about to let Rome sin twice against philosophy. Cf. E. Gilson, *Discours de la méthode: texte et commentaire*, pp. 439–442.

own satisfaction, in regard to certain problems that pertain to the speculative sciences or my attempt at governing my moral conduct by means of the reasons which the method taught me, I believed I was under no obligation to write anything. For as to moral conduct, each person agrees so much with his own opinion, that one could find as many reformers as heads, were it permitted for those other than the ones God has established as rulers over his peoples or to whom he has given sufficient grace and zeal to be prophets to try to change anything; although my speculations pleased me very much, I had believed the others also had their speculations which perhaps pleased them even more. But as soon as I had acquired some general notions in the area of physics, and, beginning to test them on various specific difficulties, I had noticed just how far they can lead and how much they differ from the principles that people have used up until the present, I believed I could not keep them hidden away without greatly sinning against the law that obliges us to procure as best we can the common good of all men. For these general notions show me that it is possible to arrive at knowledge that is very useful in life and that in place of the speculative philosophy taught in the Schools, one can find a *62* practical one, by which, knowing the force and the actions of fire, water, air, stars, the heavens, and all the other bodies that surround us, just as we understand the various skills of our craftsmen, we could, in the same way, use these objects for all the purposes for which they are appropriate, and thus make ourselves, as it were, masters and possessors of nature. This is desirable not only for the invention of an infinity of devices that would enable us to enjoy without pain the fruits of the earth and all the goods one finds in it, but also principally for the maintenance of health, which unquestionably is the first good and the foundation of all the other goods in this life; for even the mind depends so greatly upon the temperament and on the disposition of the organs of the body that, were it possible to find some means to make men generally more wise and competent than they have been up until now, I believe that one should look to medicine to find this means. It is true that the medicine currently practiced contains little of such usefulness; but without trying to ridicule it, I am sure that there is no one, not even among those in the medical profession who would not admit that everything we know is almost nothing in comparison to what remains to be known, and that we might rid ourselves of an infinity of maladies, both of body and mind, and even perhaps also the enfeeblement brought on by old age, were one to have a sufficient knowledge of their causes and of all the remedies that nature has provided us. For, desiring to spend my entire life searching for so *63* needed a science, and having found a road that seems to me such that, by following it, one ought infallibly to find that science, were it not for the fact that one is stopped either by the brevity of life or by a lack of experience, I judged there to be no better remedy against these two impediments than thus to convey faithfully to the public what little I had found and to urge good minds to try to advance beyond this, by contributing, each according to his inclination and ability, to the experiments one must conduct and also by conveying to the public everything they learned, so that later inquirers, beginning where their predecessors had left off, and thus, in joining together the lives and works of many, we might all together advance much further than a single individual could on his own.

Moreover, I noticed, in regard to experiments, that they become more necessary as one becomes more advanced in knowledge. For in the beginning it is better to make use only of what presents itself to our senses of its own accord and which we could not ignore, provided we reflect just a little on it, than to search for unusual and contrived experiments. The reason is that the most unusual ones often deceive one when one does not know yet the causes of the most ordinary experiments, and that the circumstances on which the unusual ones depend are almost always so specific and minute that it is very difficult to observe them. But the order I have held to has been the following. First, I tried to find in a general *64* way the principles or first causes of all that is or can be in the world, but not considering anything to this end except God alone who created the world, and not drawing these principles from any other source but from certain seeds of truth that are in our souls. After this I examined which ones were the first and most ordinary effects that could be deduced from these causes; it seemed to me that I had thus found the heavens, stars, an earth, and even, on the earth, water, air, fire, minerals, and other things that are the most common of all and the simplest—and hence the easiest to know. Then, when I wanted to descend to the more particular ones, so many different ones were presented to me that I did not believe it possible for the human mind to distinguish the forms or species of bodies that are on the earth from an infinity of others that could have been—had it been the will of God to have put them there—or, as a consequence, to make them serviceable to us, unless one goes ahead to causes through effects and makes use of many particular experiments. After this, passing my mind again over all the objects that ever presented themselves to my senses, I dare say that I have never seen anything that I could not explain with sufficient ease through the principles I have found. But it is also necessary for me to admit that the power of nature is so ample and so vast, and that these principles are so simple and so general, that I observe almost no specific effect without my first knowing that it can be deduced *65* in many different ways, and that my greatest difficulty is ordinarily to find in which of these ways the effect actually depends. For, to this end, I know of no other expedient except to search once more for some experiments that are such that their outcomes are not the same, if it is in one of these ways rather than in another that one ought to explain the effect. As to the rest, I am now at the point where, it seems to me, I see quite well in what direction one must go in order to do the majority of the experiments that can serve this purpose; but I also see that they are of such a nature and in such a multitude that neither my hands nor my financial resources, even if I had a thousand times more than I have, would be sufficient for all of them; so that, according as I henceforth have the wherewithal to do more or less of them, I shall also more or less go forward in the knowledge of nature. I promised myself to make this known through the treatise I have written, and to show in it so clearly the utility that the public can gain from it; I urge all those who desire the general well being of men, that is, all those who really are virtuous and not through false pretenses or merely through reputation, to communicate those experiments that they have already done as well as to help me in the search for those that remain to be done.

But since then I have had other reasons that have made me change my mind and to believe that I really ought to continue to write about all the things I judged of any importance as soon as I discovered the truth with respect to them, and to take the same care as I would if I wanted to print them. I did this as much to have more of an occasion to examine them *66* carefully (since without doubt one always looks more carefully at what one believes will be seen by many than at what one does only for oneself; and often the things that seemed to me to be true when I began to conceive them have appeared false to me when I have decided to put them on paper), as not to lose the occasion to benefit the public, if I am up to it, and so that, if my writings are worth anything, those who will have them after my death can use them in the way that would be most fitting; but that I should absolutely refuse to have them published during my lifetime, so that neither the hostilities and the controversies to which they might be subject, nor even such reputation as they might gain for me, would give me any occasion for wasting the time I have intended to use for self-instruction. For although it is true that each man is obliged to see as best he can to the good of others, and that being useful to no one is actually to be worthless, still it is also true that our concern ought to extend further than the present, and that it is well to omit things that perhaps would yield a profit to those who are living, when it is one's purpose to do other things that yield even more profit to our posterity. And I really want people to understand that what little I have learned up until now is almost nothing in comparison to what I do not know and to what I do not despair of being able to know; for it is almost the same with those who little by little discover *67* the truth in the sciences, as it is with those who, upon beginning to become wealthy, have less trouble in making large acquisitions than they had had before, when they were poorer, in making very small ones. Or one might well compare them to the leaders of an army whose forces regularly grow in proportion to their victories; they need more leadership to maintain themselves after the loss of a battle than they do after they have succeeded in taking cities or provinces. For one truly engages in battles when one tries to overcome all the difficulties and mistakes that keep us from arriving at the knowledge of the truth; and it is really a loss of a battle to accept a false opinion touching on a rather general and important matter, and it requires afterward much more skill to recover one's earlier position than to make great progress when one already has principles that are certain. For myself, if I have already found any truths in the sciences (and I hope the things contained in this volume will cause one to judge that I have found some), I can say that these are only the results and offshoots of five or six major difficulties that I have surmounted; and I count them as so many battles in which luck was on my side. I will not even be fearful of saying that I think I only need to win two or three others like them and I shall have entirely achieved the completion of my plans, and that my age is not so advanced that, according to the usual course of nature, I might *68* not still have enough leisure to bring this about. But I believe I am all the more obligated to manage well the time remaining to me, the more hope I have of being able to use it well; doubtless I had many opportunities to waste it, had I published the foundations of my physics. For although they are nearly all so evident that it is necessary only to understand them in order to believe them, and although there has never been one for which I did not believe I could give demonstrations, nevertheless, since it is impossible for them to be in agreement with all the various opinions of other men, I foresee that I would often be distracted by the hostilities these matters would engender.

One could say that these hostilities might be useful both to make me aware of my faults, as well as, were I to have anything worthwhile, to make in this way others more knowledgeable about it, and, since many can see more than one man alone, that, beginning right now to use my method, others might help me also with their inventions. But, although I acknowledge that I am extremely prone to err and that I almost never rely on the first thoughts that come to me, still the experience I have of the objections one might make against me stops me from hoping for any profit from all this. For I have already often met with the judgments both of those I took to be my friends, as well as of various others whom I took to be indifferent, and even of those too whose maliciousness and envy I knew would try very hard to discover what affection would hide from my friends. It is rare that anyone has raised an objection to me that I had not foreseen at all, unless it were very far removed from my subject; thus I have almost never found *69* any critic of my opinions who did not seem either less rigorous or less objective than myself. And I have never observed that, through the method of disputation practiced in the Schools, any truth was discovered that had until then been unknown. For, while each person in the dispute tries to win, he is more concerned with putting on a good show than with weighing the arguments on both sides; and those who long have been strong advocates are not, on that account, better judges later.

As to the utility that others would receive from the communication of my thoughts, it cannot be so terribly great, given that I have not yet taken them so far that there is not any need to add many things before applying them to common use. And I believe I can say without conceit that, if there is anyone who can do this, it ought to be me rather than someone else: not that there cannot be in the world many minds incomparably greater than my own, but that one cannot conceive a thing so well and make it one's own when one learns it from another as one can when one discovers it for oneself. This is so true in this matter that, although I have often explained some of my opinions to people with great minds who, while I spoke to them, seemed to understand these opinions quite well, nevertheless, when they repeated them, I noticed that they had almost always altered them in such a way that I could no longer acknowledge them to be mine. At this time I am very happy to ask our posterity never to believe *70* the things people say came from me, unless I myself have revealed them. And I am not at all surprised at the extravagances attributed to all those ancient philosophers whose writings we do not have; on that account I do not judge that their thoughts were terribly unreasonable, given that they were the greatest minds of their time, but only that someone has given us a bad account of them. For one sees also that their followers almost never surpassed them; and I am sure that the most impassioned of those who now follow Aristotle would believe themselves fortunate, were they to have as much knowledge of nature as he had, even if a condition of this were that they could never have any more knowledge. They are like ivy that tends to climb no higher than the trees supporting it, and even which often tends downward again after it has reached the top; for it seems to me also that those in the Schools also tend downward again, that is, they make themselves somehow less learned than had they abstained from studying, who, being unhappy with knowing only what is intelligibly explained in their author, also desire to find the solutions to many difficulties about which he has said nothing and about which he has never thought. Still, their manner of philosophizing is very congenial to those who have only very mediocre minds, for the obscurity of the distinctions and the principles they use is the reason why they can speak as boldly of all those things *71* as if they know them and why they can maintain everything they say against the subtlest and most capable, without there being any way of convincing them. In this they seem to me like a blind man, who, to fight without a handicap against someone who is sighted, makes his opponent go into the depths of a very dark cave, and I can say that they have an interest in my abstaining from publishing the principles of the philosophy I use; for my principles being very simple and very evident, I would, by publishing them, be doing almost the same as if I were to open some windows and make some light of day enter that cave where they have descended to fight. But even the greatest minds have no reason for wanting to know them; for if they want to know how to speak about everything and to acquire the reputation for being learned, they will achieve their goal more easily by contenting themselves with the appearance of truth, which can be found without much difficulty in all sorts of matters, rather than by seeking the truth, which can only be discovered little by little in a few matters and which, when it is a question of speaking about other matters, obliges one to confess frankly that he does not know them. But if they prefer the knowledge of a few truths to the vanity of appearing to be ignorant of nothing, as undoubtedly is preferable, and if they desire to follow a plan similar to my own, they do not, on this account, need me to say anything more except what I have already said in this discourse. For, if they are able to pass further than I have done, they are also all the more able to find for themselves everything that I think I have found. Given that, having examined everything in an orderly way, it is certain that what still remains *72* for me to discover is in itself more difficult and more hidden than what I have found up to this time; and they would take much less pleasure in learning it from me than from themselves. Moreover, the habit they will acquire of seeking first the easy things, then passing gradually by degrees to more difficult ones, will serve them better than all my instructions could do. As for myself, I am convinced that, had someone taught me from my youth all the truths for which I have sought demonstrations, and had I had no difficulty in learning them, I might perhaps have never learned any other truths, and at least I would never have acquired the habit and faculty I think I have for finding new truths, to the extent I apply myself in searching for them. And, in a word, if there is in the world any task that cannot be finished by anyone but the person who began it, it is that on which I am now working.

It is true that, with respect to experiments that can help here, one man alone cannot suffice to do them all, but he cannot as profitably use hands other than his own, unless they be those of craftsmen, or of such people as he can pay and for whom the hope of gain (which is a very effective means) would make do precisely what he ordered them to do. For, as to volunteers, who, out of curiosity or a desire to learn, offer themselves in order perhaps to help him, aside from usually being high on promises and low on performance and from having grand ideas none of which will come to anything, they inevitably want to be paid by an explanation of various *73* difficulties, or at least by compliments or by useless conversations which could not cost him so little of his time that it would not be a loss to him. And as to the experiments that others have already done, even when these people would desire to communicate them to him (what those who regard them as their secrets never do), they are for the most part composed of so many details and superfluities that it would be very hard to discern the truth in them; besides, one finds almost all of them to be so badly explained or even so false, since those who have done them force themselves to make them appear in conformity with their principles, that, had there been some experiments they might use, they cannot be worth the time one would have to spend in choosing them. Thus if there were in the world anyone whom one knows with certainty to be capable of finding the greatest and most beneficial things possible, and for this reason the other men fully exerted themselves to help him succeed in his plans, I do not see that they could do a thing for him except to make a donation toward the expense of the experiments he needs and, for the rest, to keep his leisure from being wasted by the importunity of anyone. But, although I do not so much presume of myself to want to promise anything out of the ordinary, nor do I feed on such vain thoughts, as to imagine that the public ought especially to be interested in my plans, still I do not have so base a soul that I wish to accept from anyone any favor that someone might think I *74* did not deserve.

All these considerations taken together were the reason why, three years ago, I did not want to unveil the treatise I had on hand, and why I had made a resolution not to make public any other treatise during my lifetime, which was so general, or one on the basis of which one could understand the foundations of my physics. But since then there have been yet again two more reasons that obliged me to attach here certain specific essays and to give the public some accounting of my actions and my plans. The first is that, if I failed to include it before, many who knew of the intention I once had to have published certain writings could imagine that the reasons why I abstain from doing so were less to my credit than was actually the case. For granted I did not love glory excessively, or even, if I dare say, that I hated it inasmuch as I judge it contrary to tranquility which I esteem above all things, still too, I had never tried to hide my actions as if they were crimes, nor had I used a lot of precaution so as not to be known; this is so both because I had believed it would do me harm as well as because it would have given me a certain disquiet that again would have been contrary to the perfect repose of mind that I seek. Since, being always thus indifferent as to whether I am known or not, I could not prevent my acquiring a certain type of reputation, I believed I ought to do my best at least to save myself from having a bad one. The other 75 reason that obliged me to write this is that, seeing more and more every day the delay that my plan of self-instruction is suffering, because of the infinity of experiments I need to make and that it is impossible to carry out without the help of someone else, although I do not flatter myself into hoping that the public will become greatly involved in my affairs, still I do not wish so much to fail myself as to give cause to those who survive me to reproach me someday for having been able to leave them many things that were much better than I had done, had I not overly neglected to make them understand how they could contribute to my plans.

And I have thought that it was easy to choose certain matters that, without being subject to a lot of controversy or obliging me to declare more of my principles than I desired, would allow me to show quite clearly what I can or cannot do in the sciences. As to this I cannot say whether I have been successful, and I do not want to prejudice the judgments of anyone in speaking for myself about my writings; but I would be very happy were a person to examine these, and, so that a person might have more of an opportunity to do this, I am asking all who have objections to make to take the trouble to send them to my publisher and, being advised about them by the publisher, I shall try to publish my reply at the same time as the objections; and by this means, seeing both of them together, readers will more easily judge the truth of the matter. For I do not promise to make long replies, but only to admit my errors very candidly, if I know *76* of any, or, if I cannot find any, to say simply what I believe is needed for the defense of what I have written—without adding an explanation of anything new—so that I am not engaging one objector after another in an endless procession.

And if any of those matters about which I spoke at the beginning of the *Dioptrics* and the *Meteorology* be shocking at first glance, because I call them "suppositions" and seem unwilling to prove them, I entreat the reader to have the patience to read the whole thing with attention; I hope he will find it satisfactory. For it seems to me that the reasons follow each other in such a way that, just as the last are demonstrated by the first—which are their causes—so these first are reciprocally demonstrated through the last—which are their effects. And one ought not to imagine that I am here committing the error logicians call a vicious circle; for, experience making the majority of these effects very certain, the causes from which I deduce these effects serve not so much to prove the effects as to explain them—but, on the other hand, the causes are what is proven by the effects. And I have called them "suppositions" only so that a person understands that I believe I can deduce them from the first truths that I have explained above. But I have expressly not wanted to do so, in order to prevent certain minds that imagine they know in one day all that someone else has thought about for twenty years as soon as he has said but two or three words to them, and who are the more subject to error and less capable of truth—the more penetrating and lively they are—from *77* being able there to take the occasion to build some extravagant philosophy on what they believe are my principles—and to prevent this from being attributed to me. For as to the opinions that are entirely my own, I do not excuse them for being new, given that, were one to consider well the reasons, I am certain that one would find them so simple and so in conformity with common sense that they seem less extraordinary and less strange than any others one could have on the same subjects. And I do not crow about being the first discoverer of any of them, but rather that I have never received them either because they have been said by others, or because they have not been, but only because reason persuades me of them.

If craftsmen cannot immediately execute the invention explained in the *Dioptrics,* I do not believe one can say, on that account, that it is bad; for, since skill and habitual disposition are needed to make and maintain the machines I have described, without any detail being overlooked, I would be no less astonished were they to succeed on the first try than were someone able to learn in one day to play the lute with distinction simply because one has been given a good musical score. And if I write in French, the language of my country, rather than in Latin, the language of my teachers, it is because I hope that those who use only their natural reason in its purity will judge my opinions, rather than those who believe only in old books. And as to those who combine good sense with study, to whom *78* alone I submit as my judges, they will not, I am certain, be so partial to Latin that they refuse to understand my reasons because I explain them in a vernacular language.

As to the rest, I do not want to speak here in specifics about the progress I hope to make in the future in the sciences, or to make any promise to the public that I am not certain of keeping; but I shall say simply that I have resolved to spend my remaining lifetime only in trying to acquire a knowledge of nature which is such that one could deduce from it rules for medicine that are more certain than those in use at present, and that my inclination puts me at such a great distance from any other sort of plan—principally those that can be useful to some person only while harmful to others—that if circumstances force me to busy myself with them, I do not believe I could succeed. On this I here declare that I know very well that it cannot help make me a man of stature in the eyes of the world, but for that matter I have no desire to be one; and I shall always try to be more obliged to those by whose favor I shall enjoy my leisure without obstacle than to those who might offer me the most honorable positions on earth.

END

Meditations on First Philosophy

NOTE ON THE TRANSLATION

The translation is based entirely on the Latin version of the *Meditations* found in volume seven of the Adam and Tannery edition of Descartes's works. It has been argued by Baillet, Descartes's early biographer, that the French "translation" by de Luynes is superior to the Latin version because it contains many additions and clarifications made by Descartes himself. However, I have not used the French version, because it contains inconsistencies and shifts that muddle more than clarify the original Latin text. The numbers found in the margins of the present translation refer to the page numbers of the Latin text in the Adam and Tannery edition.

In one instance, I found that the Latin text did not square with Descartes's clear intention. A footnote conveys my suggestion as to Descartes's actual intention in the passage.

D.A.C.

To those Most Wise and Distinguished Men,
the Dean and Doctors of the Faculty of Sacred Theology of Paris
René Descartes Sends Greetings

1

So right is the cause that impels me to offer this work to you, that I am confident you too will find it equally right and thus take up its defense, once you have understood the plan of my undertaking; so much is this the case that I have no better means of commending it here than to state briefly what I have sought to achieve in this work.

I have always thought that two issues—namely, God and the soul—are chief among those that ought to be demonstrated with the aid of philosophy rather than theology. For although it suffices for us believers to believe by faith that the human soul does not die with the body, and that God exists, certainly no unbelievers seem capable of being persuaded of any religion or even of almost any moral virtue, until these two are first proven to them by natural reason. And since in this life greater rewards are often granted to vices than to virtues, few would prefer what is right to what is useful, if they neither feared God nor anticipated an afterlife. Granted, it is altogether true that we must believe in God's existence because it is taught in the Holy Scriptures, and, conversely, that we must believe the Holy Scriptures because they have come from God. This is because, of course, since faith is a gift from God, the very same one who gives the grace that is necessary for believing the rest can also give the grace to believe that he exists. Nonetheless, this reasoning cannot be proposed to unbelievers because they would judge it to be circular. In fact, I have observed that not only do you and all other theologians affirm that one can prove the existence of God by natural reason, but also that one may infer from Sacred Scripture that the knowledge of him is easier to achieve than the many things we know about creatures, and is so utterly easy that those without this knowledge are blameworthy. For this is clear from *Wisdom*, Chapter 13, where it is said: "They are not to be excused, for if their capacity for knowing were so great that they could think well of this world, how is it that they did not find the Lord of it even more easily?" And in *Romans*, Chapter 1, it is said that they are "without excuse." And again in the same passage it appears we are being warned with the words: "What is known of God is manifest in them," that everything that can be known about God can be shown by reasons drawn exclusively from our own mind. For this reason, I did not think it unbecoming for me to inquire how this

2

may be the case, and by what path God may be known more easily and with greater certainty than the things of this world.

3 And as to the soul, there are many who have regarded its nature as something into which one cannot easily inquire, and some have even gone so far as to say that human reasoning convinces them that the soul dies with the body, while it is by faith alone that they hold the contrary position. Nevertheless, because the Lateran Council held under Leo X, in Session 8, condemned such people and expressly enjoined Christian philosophers to refute their arguments and to use all their powers to demonstrate the truth, I have not hesitated to undertake this task as well.

Moreover, I know that there are many irreligious people who refuse to believe that God exists and that the human mind is distinct from the body—for no other reason than their claim that up until now no one has been able to demonstrate these two things. By no means am I in agreement with these people; on the contrary, I believe that nearly all the arguments which have been brought to bear on these questions by great men have the force of a demonstration, when they are adequately understood, and I am convinced that hardly any arguments can be given that have not already been discovered by others. Nevertheless, I judge that there is no greater task to perform in philosophy than assiduously to seek out, once and for all, the best of all these arguments and to lay them out so precisely and plainly that henceforth all will take them to be true demonstrations. And finally, I was strongly urged to do this by some people who knew that I had developed a method for solving all sorts of problems in the sciences—not a new one, mind you, since nothing is more ancient than the truth, but one they had seen me use with some success in other areas. Accordingly, I took it to be my task to attempt something on this subject.

4 This treatise contains all that I have been able to accomplish. Not that I have attempted to gather together in it all the various arguments that could be brought forward as proof of the very same conclusions, for this does not seem worthwhile, except where no one proof is sufficiently certain. Rather, I have sought out the primary and chief arguments, so that I now make bold to propose these as most certain and evident demonstrations. Moreover, I will say in addition that these arguments are such that I believe there is no way open to the human mind whereby better ones could ever be found. For the urgency of the cause, as well as the glory of God, to which this entire enterprise is referred, compels me here to speak somewhat more freely on my own behalf than is my custom. But although I believe these arguments to be certain and evident, still I am not thereby convinced that they are suited to everyone's grasp. In geometry there are many arguments developed by Archimedes, Apollonius, Pappus, and others, which are taken by everyone to be evident and certain because they contain absolutely nothing which, considered by itself, is not quite easily known, and in which what follows does not square exactly with what has come before. Nevertheless they are rather lengthy and require a particularly attentive reader; thus only a small handful of people understand them. Likewise, although the arguments I use here do, in my opinion, equal or even surpass those of geometry in certitude and obviousness, nevertheless I am fearful that many people will not be capable of adequately perceiving them, both because they too are a bit lengthy, with some of them depending on still others, and also because, first and foremost, they demand a mind that is quite free from prejudices and that can easily withdraw itself from association with the senses. Certainly there are not to be found in the world more people with an aptitude for metaphysical studies than those with an aptitude for geometry. Moreover, there is the *5* difference that in geometry everyone is of a mind that usually nothing is put down in writing without there being a sound demonstration for it; thus the inexperienced more frequently err on the side of assenting to what is false, wanting as they do to give the appearance of understanding it, than on the side of denying what is true. But it is the reverse in philosophy: since it is believed that there is no issue that cannot be defended from either side, few look for the truth, and many more prowl about for a reputation for profundity by arrogantly challenging whichever arguments are the best.

And therefore, regardless of the force of my arguments, because they are of a philosophical nature I do not anticipate that what I will have accomplished through them will be very worthwhile unless you assist me with your patronage. Your faculty is held in such high esteem in the minds of all, and the name of the Sorbonne has such authority, that not only in matters of faith has no association, with the exception of the councils of the Church, been held in such high regard as yours, but even in human philosophy nowhere is there thought to be greater insightfulness and solidity, or greater integrity and wisdom in rendering judgments. Should you deign to show any interest in this work, I do not doubt that, first of all, its errors would be corrected by you (for I am mindful not only of my humanity but also, and most especially, of my ignorance, and thus do not claim that there are no errors in it); second, what is lacking would be added, or what is not sufficiently complete would be perfected, or what is in need of further discussion would be expanded upon more fully, either by yourselves or at least by me, after you have given me your guidance;

and finally, after the arguments contained in this work proving that God exists and that the mind is distinct from the body have been brought (as 6 I am confident they can be) to such a level of lucidity that these arguments ought to be regarded as the most precise of demonstrations, you may be of a mind to make such a declaration and publicly attest to it. Indeed, should this come to pass, I have no doubt that all the errors that have ever been entertained regarding these issues would shortly be erased from the minds of men. For the truth itself will easily cause other men of intelligence and learning to subscribe to your judgment. Your authority will cause the atheists, who more often than not are dilettantes rather than men of intelligence and learning, to put aside their spirit of contrariness, and perhaps even to defend the arguments which they will come to know are regarded as demonstrations by all who are discerning, lest they appear not to understand them. And finally, everyone else will readily give credence to so many indications of support, and there will no longer be anyone in the world who would dare call into doubt either the existence of God or the real distinction between the soul and the body. Just how great the usefulness of this thing might be, you yourselves, in virtue of your singular wisdom, are in the best position of anyone to judge; nor would it behoove me to commend the cause of God and religion at any greater length to you, who have always been the greatest pillar of the Catholic Church.

I have already touched briefly on the issues of God and the human mind in my *Discourse on the Method of Rightly Conducting One's Reason and Searching for Truth in the Sciences*, published in French in 1637. The intent there was not to provide a precise treatment of them, but only to offer a sample and to learn from the opinions of readers how these issues should be treated in the future. For they seemed to me to be so important that I judged they ought to be dealt with more than once. And the path I follow in order to explain them is so little trodden and so far removed from the one commonly taken that I did not think it useful to hold forth at greater length in a work written in French and designed to be read indiscriminately by everyone, lest weaker minds be in a position to think that they too ought to set out on this path.

In the *Discourse* I asked everyone who might find something in my writings worthy of refutation to do me the favor of making me aware of it. As for what I touched on regarding these issues, only two objections were worth noting, and I will respond briefly to them here before undertaking a more precise explanation of them.

The first is that, from the fact that the human mind, when turned in on 8 itself, does not perceive itself to be anything other than a thinking thing, it does not follow that its nature or *essence* consists only in its being a thinking thing, such that the word *only* excludes everything else that also could perhaps be said to belong to the nature of the soul. To this objection I answer that in that passage I did not intend my exclusion of those things to reflect the order of the truth of the matter (I was not dealing with it then), but merely the order of my perception. Thus what I had in mind was that I was aware of absolutely nothing that I knew belonged to my essence, save that I was a thinking thing, that is, a thing having within itself the faculty of thinking. Later on, however, I will show how it follows, from the fact that I know of nothing else belonging to my essence, that nothing else really does belong to it.

The second objection is that it does not follow from the fact that I have within me an idea of a thing more perfect than me, that this idea is itself more perfect than me, and still less that what is represented by this idea exists. But I answer that there is an equivocation here in the word "idea." For "idea" can be taken either materially, for an operation of the intellect (in which case it cannot be said to be more perfect than me), or objectively, for the thing represented by means of that operation. This thing, even if it is not presumed to exist outside the intellect, can nevertheless be more perfect than me by reason of its essence. I will explain in detail in the ensuing remarks how, from the mere fact that there is within me an idea of something more perfect than me, it follows that this thing really exists.

In addition, I have seen two rather lengthy treatises, but these works, utilizing as they do arguments drawn from atheist commonplaces, focused 9 their attack not so much on my arguments regarding these issues, as on my conclusions. Moreover, arguments of this type exercise no influence over those who understand my arguments, and the judgments of many people are so preposterous and feeble that they are more likely to be persuaded by the first opinions to come along, however false and contrary to reason they may be, than by a true and firm refutation of them which they hear subsequently. Accordingly, I have no desire to respond here to these objections, lest I first have to state what they are. I will only say in general that all the objections typically bandied about by the atheists to assail the existence of God always depend either on ascribing human emotions to God, or on arrogantly claiming for our minds such power and wisdom that we attempt to determine and grasp fully what God can and ought to do. Hence these objections will cause us no difficulty, provided we but remember that our minds are to be regarded as finite, while God is to be regarded as incomprehensible and infinite.

But now, after having, to some degree, conducted an initial review of the judgments of men, here I begin once more to treat the same questions about God and the human mind, together with the starting points of the whole of first philosophy, but not in a way that causes me to have any expectation of widespread approval or a large readership. On the contrary, I do not advise anyone to read these things except those who have both the ability and the desire to meditate seriously with me, and to withdraw their minds from the senses as well as from all prejudices. I know all too well that such people are few and far between. As to those who do not take the time to grasp the order and linkage of my arguments, but will be 10 eager to fuss over statements taken out of context (as is the custom for many), they will derive little benefit from reading this work. Although perhaps they might find an occasion for quibbling in several places, still they will not find it easy to raise an objection that is either compelling or worthy of response.

But because I do not promise to satisfy even the others on all counts the first time around, and because I do not arrogantly claim for myself so much that I believe myself capable of anticipating all the difficulties that will occur to someone, I will first of all narrate in the *Meditations* the very thoughts by means of which I seem to have arrived at a certain and evident knowledge of the truth, so that I may determine whether the same arguments that persuaded me can be useful in persuading others. Next, I will reply to the objections of a number of very gifted and learned gentlemen, to whom these *Meditations* were forwarded for their examination prior to their being sent to press. For their objections were so many and varied that I have dared to hope that nothing will readily occur to anyone, at least nothing of importance, which has not already been touched upon by these gentlemen. And thus I earnestly entreat the readers not to form a judgment regarding the *Meditations* until they have deigned to read all these objections and the replies I have made to them.

In the First Meditation the reasons are given why we can doubt all things, especially material things, so long, that is, as, of course, we have no other foundations for the sciences than the ones which we have had up until now. Although the utility of so extensive a doubt is not readily apparent, nevertheless its greatest utility lies in freeing us of all prejudices, in preparing the easiest way for us to withdraw the mind from the senses, and finally, in making it impossible for to us doubt any further those things that we later discover to be true.

In the Second Meditation the mind, through the exercise of its own freedom, supposes the nonexistence of all those things about whose existence it can have even the least doubt. In so doing the mind realizes that it is impossible for it not to exist during this time. This too is of the greatest utility, since by means of it the mind easily distinguishes what things belong to it, that is, to an intellectual nature, from what things belong to the body. But because some people will perhaps expect to see proofs for the immortality of the soul in this Meditation, I think they should be put 13 on notice here that I have attempted to write only what I have carefully demonstrated. Therefore the only order I could follow was the one typically used by geometers, which is to lay out everything on which a given proposition depends, before concluding anything about it. But the first and principal prerequisite for knowing that the soul is immortal is that we form a concept of the soul that is as lucid as possible and utterly distinct from every concept of a body. This is what has been done here. Moreover, there is the additional requirement that we know that everything that we clearly and distinctly understand is true, in exactly the manner in which we understand it; however, this could not have been proven prior to the Fourth Meditation. Moreover, we must have a distinct concept of corporeal nature, and this is formulated partly in the Second Meditation itself, and partly in the Fifth and Sixth Meditations. From all this one ought to conclude that all the things we clearly and distinctly conceive as different substances truly are substances that are really distinct from one another. (This, for example, is how mind and body are conceived.) This conclusion is arrived at in the Sixth Meditation. This same conclusion is also confirmed in this Meditation in virtue of the fact that we cannot understand a body to be anything but divisible, whereas we cannot understand the mind to be anything but indivisible. For we cannot conceive of half of a mind, as we can half of any body whatever, no matter how small. From this we are prompted to acknowledge that the natures of mind and body not only are different from one another, but even, in a manner of speaking, are contraries of one another. However, I have not written any further on the matter in this work, both because these considerations suffice for showing that the annihilation of the mind does not follow from the decaying of the body (and thus these considerations suffice for giving mortals hope in an afterlife), and also because the premises from which the immortality of the mind can be inferred depend upon an account of the whole of physics. First, we need to know that absolutely all substances, that is, *14* things that must be created by God in order to exist, are by their very nature incorruptible, and can never cease to exist, unless, by the same God's denying his concurrence to them, they be reduced to nothingness. Second, we need to realize that body, taken in a general sense, is a substance and hence it too can never perish. But the human body, insofar as it differs from other bodies, is composed of merely a certain configuration of members, together with other accidents of the same sort. But the human mind is not likewise composed of any accidents, but is a pure substance. For even if all its accidents were changed, so that it understands different things, wills different things, senses different things, and so on, the mind itself does not on that score become something different. On the other hand, the human body does become something different, merely as a result of the fact that a change in the shape of some of its parts has taken place. It follows from these considerations that a body can very easily perish, whereas the mind by its nature is immortal.

In the Third Meditation I have explained at sufficient length, it seems to me, my principal argument for proving the existence of God. Nevertheless, since my intent was to draw the minds of readers as far as possible from the senses, I had no desire to draw upon comparisons based upon corporeal things. Thus many obscurities may perhaps have remained; but these, I trust, will later be entirely removed in my Replies to the Objections. One such point of contention, among others, is the following: how can the idea that is in us of a supremely perfect being have so much objective reality that it can only come from a supremely perfect cause? This is illustrated in the Replies by a comparison with a very perfect machine, the idea of which is in the mind of some craftsman. For, just as the objective ingeniousness of this idea ought to have some cause (say, the knowledge possessed by the craftsman or by someone else from whom he received this knowledge), so too, the idea of God which is in us must have God *15* himself as its cause.

In the Fourth Meditation it is proved that all that we clearly and distinctly perceive is true, and it is also explained what constitutes the nature of falsity. These things necessarily need to be known both to confirm what has preceded as well as to help readers understand what remains. (But here one should meanwhile bear in mind that in that Meditation there is no discussion whatsoever of sin, that is, the error committed in the pursuit of good and evil, but only the error that occurs in discriminating between what is true and what is false. Nor is there an examination of those matters pertaining to the faith or to the conduct of life, but merely of speculative truths known exclusively by means of the light of nature.)

In the Fifth Meditation, in addition to an explanation of corporeal nature in general, the existence of God is also demonstrated by means of a new proof. But again several difficulties may arise here; however, these are resolved later in my Replies to the Objections. Finally, it is shown how it is true that the certainty of even geometrical demonstrations depends upon the knowledge of God.

Finally, in the Sixth Meditation the understanding is distinguished from the imagination and the marks of this distinction are described. The mind is proved to be really distinct from the body, even though the mind is shown to be so closely joined to the body that it forms a single unit with it. All the errors commonly arising from the senses are reviewed; an account of the ways in which these errors can be avoided is provided. Finally, all the arguments on the basis of which we may infer the existence of material things are presented—not because I believed them to be very *16* useful for proving what they prove, namely, that there really is a world, that men have bodies, and the like (things which no one of sound mind has ever seriously doubted), but rather because, through a consideration of these arguments, one realizes that they are neither so firm nor so evident as the arguments leading us to the knowledge of our mind and of God, so that, of all the things that can be known by the human mind, these latter are the most certain and the most evident. Proving this one thing was for me the goal of these Meditations. For this reason I will not review here the various issues that are also to be treated in these Meditations as the situation arises.

Meditations
on First Philosophy
In Which
the Existence of God
and the Distinction between the Soul
and the Body
Are Demonstrated

17

MEDITATIONS ON FIRST PHILOSOPHY IN WHICH THE EXISTENCE OF GOD AND THE DISTINCTION BETWEEN THE SOUL AND THE BODY ARE DEMONSTRATED

Meditation One: Concerning Those Things That Can Be Called into Doubt

Several years have now passed since I first realized how numerous were the false opinions that in my youth I had taken to be true, and thus how doubtful were all those that I had subsequently built upon them. And thus I realized that once in my life I had to raze everything to the ground and begin again from the original foundations, if I wanted to establish anything firm and lasting in the sciences. But the task seemed enormous, and I was waiting until I reached a point in my life that was so timely that no more suitable time for undertaking these plans of action would come to pass. For this reason, I procrastinated for so long that I would henceforth be at fault, were I to waste the time that remains for carrying out the project by brooding over it. Accordingly, I have today suitably freed my mind of all cares, secured for myself a period of leisurely tranquillity, and am 18 withdrawing into solitude. At last I will apply myself earnestly and unreservedly to this general demolition of my opinions.

Yet to bring this about I will not need to show that all my opinions are false, which is perhaps something I could never accomplish. But reason now persuades me that I should withhold my assent no less carefully from opinions that are not completely certain and indubitable than I would from those that are patently false. For this reason, it will suffice for the rejection of all of these opinions, if I find in each of them some reason for doubt.

Nor therefore need I survey each opinion individually, a task that would be endless. Rather, because undermining the foundations will cause whatever has been built upon them to crumble of its own accord, I will attack straightaway those principles which supported everything I once believed.

Surely whatever I had admitted until now as most true I received either from the senses or through the senses. However, I have noticed that the senses are sometimes deceptive; and it is a mark of prudence never to place our complete trust in those who have deceived us even once.

But perhaps, even though the senses do sometimes deceive us when it is a question of very small and distant things, still there are many other matters concerning which one simply cannot doubt, even though they are derived from the very same senses: for example, that I am sitting here next to the fire, wearing my winter dressing gown, that I am holding this sheet of paper in my hands, and the like. But on what grounds could one deny that these hands and this entire body are mine? Unless perhaps I were to *19* liken myself to the insane, whose brains are impaired by such an unrelenting vapor of black bile that they steadfastly insist that they are kings when they are utter paupers, or that they are arrayed in purple robes when they are naked, or that they have heads made of clay, or that they are gourds, or that they are made of glass. But such people are mad, and I would appear no less mad, were I to take their behavior as an example for myself.

This would all be well and good, were I not a man who is accustomed to sleeping at night, and to experiencing in my dreams the very same things, or now and then even less plausible ones, as these insane people do when they are awake. How often does my evening slumber persuade me of such ordinary things as these: that I am here, clothed in my dressing gown, seated next to the fireplace—when in fact I am lying undressed in bed! But right now my eyes are certainly wide awake when I gaze upon this sheet of paper. This head which I am shaking is not heavy with sleep. I extend this hand consciously and deliberately, and I feel it. Such things would not be so distinct for someone who is asleep. As if I did not recall having been deceived on other occasions even by similar thoughts in my dreams! As I consider these matters more carefully, I see so plainly that there are no definitive signs by which to distinguish being awake from being asleep. As a result, I am becoming quite dizzy, and this dizziness nearly convinces me that I am asleep.

Let us assume then, for the sake of argument, that we are dreaming and that such particulars as these are not true: that we are opening our eyes, moving our head, and extending our hands. Perhaps we do not even have such hands, or any such body at all. Nevertheless, it surely must be admitted that the things seen during slumber are, as it were, like painted images, which could only have been produced in the likeness of true things, and that therefore at least these general things—eyes, head, hands, and the whole body—are not imaginary things, but are true and exist. For indeed when painters themselves wish to represent sirens and satyrs by *20* means of especially bizarre forms, they surely cannot assign to them utterly new natures. Rather, they simply fuse together the members of various animals. Or if perhaps they concoct something so utterly novel that nothing like it has ever been seen before (and thus is something utterly fictitious and false), yet certainly at the very least the colors from which they fashion it ought to be true. And by the same token, although even these general things—eyes, head, hands and the like—could be imaginary, still one has to admit that at least certain other things that are even more simple and universal are true. It is from these components, as if from true colors, that all those images of things that are in our thought are fashioned, be they true or false.

This class of things appears to include corporeal nature in general, together with its extension; the shape of extended things; their quantity, that is, their size and number; as well as the place where they exist; the time through which they endure, and the like.

Thus it is not improper to conclude from this that physics, astronomy, medicine, and all the other disciplines that are dependent upon the consideration of composite things are doubtful, and that, on the other hand, arithmetic, geometry, and other such disciplines, which treat of nothing but the simplest and most general things and which are indifferent as to whether these things do or do not in fact exist, contain something certain and indubitable. For whether I am awake or asleep, two plus three make five, and a square does not have more than four sides. It does not seem possible that such obvious truths should be subject to the suspicion of being false.

Be that as it may, there is fixed in my mind a certain opinion of long *21* standing, namely that there exists a God who is able to do anything and by whom I, such as I am, have been created. How do I know that he did not bring it about that there is no earth at all, no heavens, no extended thing, no shape, no size, no place, and yet bringing it about that all these things appear to me to exist precisely as they do now? Moreover, since I judge that others sometimes make mistakes in matters that they believe they know most perfectly, may I not, in like fashion, be deceived every time I add two and three or count the sides of a square, or perform an even simpler operation, if that can be imagined? But perhaps God has not willed that I be deceived in this way, for he is said to be supremely good. Nonetheless, if it were repugnant to his goodness to have created me such that I be deceived all the time, it would also seem foreign to that same goodness to permit me to be deceived even occasionally. But we cannot make this last assertion.

Perhaps there are some who would rather deny so a powerful a God than believe that everything else is uncertain. Let us not oppose them; rather, let us grant that everything said here about God is fictitious. Now they suppose that I came to be what I am either by fate, or by chance, or by a connected chain of events, or by some other way. But because being deceived and being mistaken appear to be a certain imperfection, the less powerful they take the author of my origin to be, the more probable it will be that I am so imperfect that I am always deceived. I have nothing to say in response to these arguments. But eventually I am forced to admit that there is nothing among the things I once believed to be true which it is not permissible to doubt—and not out of frivolity or lack of forethought, but for valid and considered reasons. Thus I must be no less careful to *22* withhold assent henceforth even from these beliefs than I would from those that are patently false, if I wish to find anything certain.

But it is not enough simply to have realized these things; I must take steps to keep myself mindful of them. For long-standing opinions keep returning, and, almost against my will, they take advantage of my credulity, as if it were bound over to them by long use and the claims of intimacy. Nor will I ever get out of the habit of assenting to them and believing in them, so long as I take them to be exactly what they are, namely, in some respects doubtful, as has just now been shown, but nevertheless highly probable, so that it is much more consonant with reason to believe them than to deny them. Hence, it seems to me I would do well to deceive myself by turning my will in completely the opposite direction and pretend for a time that these opinions are wholly false and imaginary, until finally, as if with prejudices weighing down each side equally, no bad habit should turn my judgment any further from the correct perception of things. For indeed I know that meanwhile there is no danger or error in following this procedure, and that it is impossible for me to indulge in too much distrust, since I am now concentrating only on knowledge, not on action.

Accordingly, I will suppose not a supremely good God, the source of truth, but rather an evil genius, supremely powerful and clever, who has directed his entire effort at deceiving me. I will regard the heavens, the air, the earth, colors, shapes, sounds, and all external things as nothing but the bedeviling hoaxes of my dreams, with which he lays snares for my credulity. I will regard myself as not having hands, or eyes, or flesh, or *23* blood, or any senses, but as nevertheless falsely believing that I possess all these things. I will remain resolute and steadfast in this meditation, and even if it is not within my power to know anything true, it certainly is within my power to take care resolutely to withhold my assent to what is false, lest this deceiver, however powerful, however clever he may be, have any effect on me. But this undertaking is arduous, and a certain laziness brings me back to my customary way of living. I am not unlike a prisoner who enjoyed an imaginary freedom during his sleep, but, when he later begins to suspect that he is dreaming, fears being awakened and nonchalantly conspires with these pleasant illusions. In just the same way, I fall back of my own accord into my old opinions, and dread being awakened, lest the toilsome wakefulness which follows upon a peaceful rest must be spent thenceforward not in the light but among the inextricable shadows of the difficulties now brought forward.

MEDITATION TWO: Concerning the Nature of the Human Mind: That It Is Better Known Than the Body

Yesterday's meditation has thrown me into such doubts that I can no longer ignore them, yet I fail to see how they are to be resolved. It is as if I had suddenly fallen into a deep whirlpool; I am so tossed about that *24* I can neither touch bottom with my foot, nor swim up to the top. Nevertheless I will work my way up and will once again attempt the same path I entered upon yesterday. I will accomplish this by putting aside everything that admits of the least doubt, as if I had discovered it to be completely false. I will stay on this course until I know something certain, or, if nothing else, until I at least know for certain that nothing is certain. Archimedes sought but one firm and immovable point in order to move the entire earth from one place to another. Just so, great things are also to be hoped for if I succeed in finding just one thing, however slight, that is certain and unshaken.

Therefore I suppose that everything I see is false. I believe that none of what my deceitful memory represents ever existed. I have no senses whatever. Body, shape, extension, movement, and place are all chimeras. What then will be true? Perhaps just the single fact that nothing is certain.

But how do I know there is not something else, over and above all those things that I have just reviewed, concerning which there is not even the slightest occasion for doubt? Is there not some God, or by whatever name I might call him, who instills these very thoughts in me? But why would

I think that, since I myself could perhaps be the author of these thoughts? Am I not then at least something? But I have already denied that I have 25 any senses and any body. Still I hesitate; for what follows from this? Am I so tied to a body and to the senses that I cannot exist without them? But I have persuaded myself that there is absolutely nothing in the world: no sky, no earth, no minds, no bodies. Is it then the case that I too do not exist? But doubtless I did exist, if I persuaded myself of something. But there is some deceiver or other who is supremely powerful and supremely sly and who is always deliberately deceiving me. Then too there is no doubt that I exist, if he is deceiving me. And let him do his best at deception, he will never bring it about that I am nothing so long as I shall think that I am something. Thus, after everything has been most carefully weighed, it must finally be established that this pronouncement "I am, I exist" is necessarily true every time I utter it or conceive it in my mind.

But I do not yet understand sufficiently what I am—I, who now necessarily exist. And so from this point on, I must be careful lest I unwittingly mistake something else for myself, and thus err in that very item of knowledge that I claim to be the most certain and evident of all. Thus, I will meditate once more on what I once believed myself to be, prior to embarking upon these thoughts. For this reason, then, I will set aside whatever can be weakened even to the slightest degree by the arguments brought forward, so that eventually all that remains is precisely nothing but what is certain and unshaken.

What then did I use to think I was? A man, of course. But what is a man? Might I not say a "rational animal"? No, because then I would have to inquire what "animal" and "rational" mean. And thus from one question I would slide into many more difficult ones. Nor do I now have enough free time that I want to waste it on subtleties of this sort. Instead, permit 26 me to focus here on what came spontaneously and naturally into my thinking whenever I pondered what I was. Now it occurred to me first that I had a face, hands, arms, and this entire mechanism of bodily members: the very same as are discerned in a corpse, and which I referred to by the name "body." It next occurred to me that I took in food, that I walked about, and that I sensed and thought various things; these actions I used to attribute to the soul. But as to what this soul might be, I either did not think about it or else I imagined it a rarified I-know-not-what, like a wind, or a fire, or ether, which had been infused into my coarser parts. But as to the body I was not in any doubt. On the contrary, I was under the impression that I knew its nature distinctly. Were I perhaps tempted to describe this nature such as I conceived it in my mind, I would have described it thus: by "body," I understand all that is capable of being bounded by some shape, of being enclosed in a place, and of filling up a space in such a way as to exclude any other body from it; of being perceived by touch, sight, hearing, taste, or smell; of being moved in several ways, not, of course, by itself, but by whatever else impinges upon it. For it was my view that the power of self-motion, and likewise of sensing or of thinking, in no way belonged to the nature of the body. Indeed I used rather to marvel that such faculties were to be found in certain bodies.

But now what am I, when I suppose that there is some supremely powerful and, if I may be permitted to say so, malicious deceiver who deliberately tries to fool me in any way he can? Can I not affirm that I possess at least a small measure of all those things which I have already said belong to the nature of the body? I focus my attention on them, I *27* think about them, I review them again, but nothing comes to mind. I am tired of repeating this to no purpose. But what about those things I ascribed to the soul? What about being nourished or moving about? Since I now do not have a body, these are surely nothing but fictions. What about sensing? Surely this too does not take place without a body; and I seemed to have sensed in my dreams many things that I later realized I did not sense. What about thinking? Here I make my discovery: thought exists; it alone cannot be separated from me. I am; I exist—this is certain. But for how long? For as long as I am thinking; for perhaps it could also come to pass that if I were to cease all thinking I would then utterly cease to exist. At this time I admit nothing that is not necessarily true. I am therefore precisely nothing but a thinking thing; that is, a mind, or intellect, or understanding, or reason—words of whose meanings I was previously ignorant. Yet I am a true thing and am truly existing; but what kind of thing? I have said it already: a thinking thing.

What else am I? I will set my imagination in motion. I am not that concatenation of members we call the human body. Neither am I even some subtle air infused into these members, nor a wind, nor a fire, nor a vapor, nor a breath, nor anything I devise for myself. For I have supposed these things to be nothing. The assumption still stands; yet nevertheless I am something. But is it perhaps the case that these very things which I take to be nothing, because they are unknown to me, nevertheless are in fact no different from that "me" that I know? This I do not know, and I will not quarrel about it now. I can make a judgment only about things that are known to me. I know that I exist; I ask now who is this "I" whom I know? Most certainly, in the strict sense the knowledge of this "I" does not depend upon things of whose existence I do not yet have knowledge. *28*

Therefore it is not dependent upon any of those things that I simulate in my imagination. But this word "simulate" warns me of my error. For I would indeed be simulating were I to "imagine" that I was something, because imagining is merely the contemplating of the shape or image of a corporeal thing. But I now know with certainty that I am and also that all these images—and, generally, everything belonging to the nature of the body—could turn out to be nothing but dreams. Once I have realized this, I would seem to be speaking no less foolishly were I to say: "I will use my imagination in order to recognize more distinctly who I am," than were I to say: "Now I surely am awake, and I see something true; but since I do not yet see it clearly enough, I will deliberately fall asleep so that my dreams might represent it to me more truly and more clearly." Thus I realize that none of what I can grasp by means of the imagination pertains to this knowledge that I have of myself. Moreover, I realize that I must be most diligent about withdrawing my mind from these things so that it can perceive its nature as distinctly as possible.

But what then am I? A thing that thinks. What is that? A thing that doubts, understands, affirms, denies, wills, refuses, and that also imagines and senses.

Indeed it is no small matter if all of these things belong to me. But why should they not belong to me? Is it not the very same "I" who now doubts almost everything, who nevertheless understands something, who affirms that this one thing is true, who denies other things, who desires to know more, who wishes not to be deceived, who imagines many things even against my will, who also notices many things which appear to come from the senses? What is there in all of this that is not every bit as true as the *29* fact that I exist—even if I am always asleep or even if my creator makes every effort to mislead me? Which of these things is distinct from my thought? Which of them can be said to be separate from myself? For it is so obvious that it is I who doubt, I who understand, and I who will, that there is nothing by which it could be explained more clearly. But indeed it is also the same "I" who imagines; for although perhaps, as I supposed before, absolutely nothing that I imagined is true, still the very power of imagining really does exist, and constitutes a part of my thought. Finally, it is this same "I" who senses or who is cognizant of bodily things as if through the senses. For example, I now see a light, I hear a noise, I feel heat. These things are false, since I am asleep. Yet I certainly do seem to see, hear, and feel warmth. This cannot be false. Properly speaking, this is what in me is called "sensing." But this, precisely so taken, is nothing other than thinking.

From these considerations I am beginning to know a little better what I am. But it still seems (and I cannot resist believing) that corporeal things—whose images are formed by thought, and which the senses themselves examine—are much more distinctly known than this mysterious "I" which does not fall within the imagination. And yet it would be strange indeed were I to grasp the very things I consider to be doubtful, unknown, and foreign to me more distinctly than what is true, what is known—than, in short, myself. But I see what is happening: my mind loves to wander and does not yet permit itself to be restricted within the confines of truth. So be it then; let us just this once allow it completely free rein, so that, a *30* little while later, when the time has come to pull in the reins, the mind may more readily permit itself to be controlled.

Let us consider those things which are commonly believed to be the most distinctly grasped of all: namely the bodies we touch and see. Not bodies in general, mind you, for these general perceptions are apt to be somewhat more confused, but one body in particular. Let us take, for instance, this piece of wax. It has been taken quite recently from the honeycomb; it has not yet lost all the honey flavor. It retains some of the scent of the flowers from which it was collected. Its color, shape, and size are manifest. It is hard and cold; it is easy to touch. If you rap on it with your knuckle it will emit a sound. In short, everything is present in it that appears needed to enable a body to be known as distinctly as possible. But notice that, as I am speaking, I am bringing it close to the fire. The remaining traces of the honey flavor are disappearing; the scent is vanishing; the color is changing; the original shape is disappearing. Its size is increasing; it is becoming liquid and hot; you can hardly touch it. And now, when you rap on it, it no longer emits any sound. Does the same wax still remain? I must confess that it does; no one denies it; no one thinks otherwise. So what was there in the wax that was so distinctly grasped? Certainly none of the aspects that I reached by means of the senses. For whatever came under the senses of taste, smell, sight, touch or hearing has now changed; and yet the wax remains.

Perhaps the wax was what I now think it is: namely that the wax itself never really was the sweetness of the honey, nor the fragrance of the flowers, nor the whiteness, nor the shape, nor the sound, but instead was a body that a short time ago manifested itself to me in these ways, and now does so in other ways. But just what precisely is this thing that I thus imagine? Let us focus our attention on this and see what remains after we *31* have removed everything that does not belong to the wax: only that it is something extended, flexible, and mutable. But what is it to be flexible and mutable? Is it what my imagination shows it to be: namely, that this piece of wax can change from a round to a square shape, or from the latter to a triangular shape? Not at all; for I grasp that the wax is capable of innumerable changes of this sort, even though I am incapable of running through these innumerable changes by using my imagination. Therefore this insight is not achieved by the faculty of imagination. What is it to be extended? Is this thing's extension also unknown? For it becomes greater in wax that is beginning to melt, greater in boiling wax, and greater still as the heat is increased. And I would not judge correctly what the wax is if I did not believe that it takes on an even greater variety of dimensions than I could ever grasp with the imagination. It remains then for me to concede that I do not grasp what this wax is through the imagination; rather, I perceive it through the mind alone. The point I am making refers to this particular piece of wax, for the case of wax in general is clearer still. But what is this piece of wax which is perceived only by the mind? Surely it is the same piece of wax that I see, touch, and imagine; in short it is the same piece of wax I took it to be from the very beginning. But I need to realize that the perception of the wax is neither a seeing, nor a touching, nor an imagining. Nor has it ever been, even though it previously seemed so; rather it is an inspection on the part of the mind alone. This inspection can be imperfect and confused, as it was before, or clear and distinct, as it is now, depending on how closely I pay attention to the things in which the piece of wax consists.

But meanwhile I marvel at how prone my mind is to errors. For although *32* I am considering these things within myself silently and without words, nevertheless I seize upon words themselves and I am nearly deceived by the ways in which people commonly speak. For we say that we see the wax itself, if it is present, and not that we judge it to be present from its color or shape. Whence I might conclude straightaway that I know the wax through the vision had by the eye, and not through an inspection on the part of the mind alone. But then were I perchance to look out my window and observe men crossing the square, I would ordinarily say I see the men themselves just as I say I see the wax. But what do I see aside from hats and clothes, which could conceal automata? Yet I judge them to be men. Thus what I thought I had seen with my eyes, I actually grasped solely with the faculty of judgment, which is in my mind.

But a person who seeks to know more than the common crowd ought to be ashamed of himself for looking for doubt in common ways of speaking. Let us then go forward and inquire when it was that I perceived more perfectly and evidently what the piece of wax was. Was it when I first saw it and believed I knew it by the external sense, or at least by the so-called common sense, that is, the power of imagination? Or do I have more perfect knowledge now, when I have diligently examined both what the wax is and how it is known? Surely it is absurd to be in doubt about this matter. For what was there in my initial perception that was distinct? What was there that any animal seemed incapable of possessing? But indeed when I distinguish the wax from its external forms, as if stripping it of its clothing, and look at the wax in its nakedness, then, even though there can be still an error in my judgment, nevertheless I cannot perceive it thus without a human mind.

But what am I to say about this mind, that is, about myself? For as yet I admit nothing else to be in me over and above the mind. What, I ask, am I who seem to perceive this wax so distinctly? Do I not know myself not only much more truly and with greater certainty, but also much more distinctly and evidently? For if I judge that the wax exists from the fact that I see it, certainly from this same fact that I see the wax it follows much more evidently that I myself exist. For it could happen that what I see is not truly wax. It could happen that I have no eyes with which to see anything. But it is utterly impossible that, while I see or think I see (I do not now distinguish these two), I who think am not something. Likewise, if I judge that the wax exists from the fact that I touch it, the same outcome will again obtain, namely that I exist. If I judge that the wax exists from the fact that I imagine it, or for any other reason, plainly the same thing follows. But what I note regarding the wax applies to everything else that is external to me. Furthermore, if my perception of the wax seemed more distinct after it became known to me not only on account of sight or touch, but on account of many reasons, one has to admit how much more distinctly I am now known to myself. For there is not a single consideration that can aid in my perception of the wax or of any other body that fails to make even more manifest the nature of my mind. But there are still so many other things in the mind itself on the basis of which my knowledge of it can be rendered more distinct that it hardly seems worth enumerating those things which emanate to it from the body. *33*

But lo and behold, I have returned on my own to where I wanted to be. For since I now know that even bodies are not, properly speaking, perceived by the senses or by the faculty of imagination, but by the intellect alone, and that they are not perceived through their being touched or seen, but only through their being understood, I manifestly know that nothing can be perceived more easily and more evidently than my own mind. But since the tendency to hang on to long-held beliefs cannot be *34* put aside so quickly, I want to stop here, so that by the length of my meditation this new knowledge may be more deeply impressed upon my memory.

MEDITATION THREE: Concerning God, That He Exists

I will now shut my eyes, stop up my ears, and withdraw all my senses. I will also blot out from my thoughts all images of corporeal things, or rather, since the latter is hardly possible, I will regard these images as empty, false and worthless. And as I converse with myself alone and look more deeply into myself, I will attempt to render myself gradually better known and more familiar to myself. I am a thing that thinks, that is to say, a thing that doubts, affirms, denies, understands a few things, is ignorant of many things, wills, refrains from willing, and also imagines and senses. For as I observed earlier, even though these things that I sense or imagine may perhaps be nothing at all outside me, nevertheless I am certain that these modes of thinking, which are cases of what I call sensing and *35* imagining, insofar as they are merely modes of thinking, do exist within me.

In these few words, I have reviewed everything I truly know, or at least what so far I have noticed that I know. Now I will ponder more carefully to see whether perhaps there may be other things belonging to me that up until now I have failed to notice. I am certain that I am a thinking thing. But do I not therefore also know what is required for me to be certain of anything? Surely in this first instance of knowledge, there is nothing but a certain clear and distinct perception of what I affirm. Yet this would hardly be enough to render me certain of the truth of a thing, if it could ever happen that something that I perceived so clearly and distinctly were false. And thus I now seem able to posit as a general rule that everything I very clearly and distinctly perceive is true.

Be that as it may, I have previously admitted many things as wholly certain and evident that nevertheless I later discovered to be doubtful. What sort of things were these? Why, the earth, the sky, the stars, and all the other things I perceived by means of the senses. But what was it about these things that I clearly perceived? Surely the fact that the ideas or thoughts of these things were hovering before my mind. But even now I do not deny that these ideas are in me. Yet there was something else I used to affirm, which, owing to my habitual tendency to believe it, I used to think was something I clearly perceived, even though I actually did not perceive it at all: namely, that certain things existed outside me, things from which those ideas proceeded and which those ideas completely resembled. But on this point I was mistaken; or rather, if my judgment was a true one, it was not the result of the force of my perception.

But what about when I considered something very simple and easy in the areas of arithmetic or geometry, for example that two plus three make five, and the like? Did I not intuit them at least clearly enough so as to affirm them as true? To be sure, I did decide later on that I must doubt these things, but that was only because it occurred to me that some God could perhaps have given me a nature such that I might be deceived even about matters that seemed most evident. But whenever this preconceived opinion about the supreme power of God occurs to me, I cannot help admitting that, were he to wish it, it would be easy for him to cause me to err even in those matters that I think I intuit as clearly as possible with the eyes of the mind. On the other hand, whenever I turn my attention to those very things that I think I perceive with such great clarity, I am so completely persuaded by them that I spontaneously blurt out these words: "let anyone who can do so deceive me; so long as I think that I am something, he will never bring it about that I am nothing. Nor will he one day make it true that I never existed, for it is true now that I do exist. Nor will he even bring it about that perhaps two plus three might equal more or less than five, or similar items in which I recognize an obvious contradiction." And certainly, because I have no reason for thinking that there is a God who is a deceiver (and of course I do not yet sufficiently know whether there even is a God), the basis for doubting, depending as it does merely on the above hypothesis, is very tenuous and, so to speak, metaphysical. But in order to remove even this basis for doubt, I should at the first opportunity inquire whether there is a God, and, if there is, whether or not he can be a deceiver. For if I am ignorant of this, it appears I am never capable of being completely certain about anything else. *36*

However, at this stage good order seems to demand that I first group all my thoughts into certain classes, and ask in which of them truth or falsity properly resides. Some of these thoughts are like images of things; to these alone does the word "idea" properly apply, as when I think of a man, or a chimera, or the sky, or an angel, or God. Again there are other thoughts that take different forms: for example, when I will, or fear, or affirm, or deny, there is always some thing that I grasp as the subject of my thought, yet I embrace in my thought something more than the likeness of that thing. Some of these thoughts are called volitions or affects, while others are called judgments. *37*

Now as far as ideas are concerned, if they are considered alone and in their own right, without being referred to something else, they cannot, properly speaking, be false. For whether it is a she-goat or a chimera that I am imagining, it is no less true that I imagine the one than the other. Moreover, we need not fear that there is falsity in the will itself or in the affects, for although I can choose evil things or even things that are utterly non-existent, I cannot conclude from this that it is untrue that I do choose these things. Thus there remain only judgments in which I must take care not to be mistaken. Now the principal and most frequent error to be found in judgments consists in the fact that I judge that the ideas which are in me are similar to or in conformity with certain things outside me. Obviously, if I were to consider these ideas merely as certain modes of my thought, and were not to refer them to anything else, they could hardly give me any subject matter for error.

Among these ideas, some appear to me to be innate, some adventitious, *38* and some produced by me. For I understand what a thing is, what truth is, what thought is, and I appear to have derived this exclusively from my very own nature. But say I am now hearing a noise, or looking at the sun, or feeling the fire; up until now I judged that these things proceeded from certain things outside me, and finally, that sirens, hippogriffs, and the like are made by me. Or perhaps I can even think of all these ideas as being adventitious, or as being innate, or as fabrications, for I have not yet clearly ascertained their true origin.

But here I must inquire particularly into those ideas that I believe to be derived from things existing outside me. Just what reason do I have for believing that these ideas resemble those things? Well, I do seem to have been so taught by nature. Moreover, I do know from experience that these ideas do not depend upon my will, nor consequently upon myself, for I often notice them even against my will. Now, for example, whether or not I will it, I feel heat. It is for this reason that I believe this feeling or idea of heat comes to me from something other than myself, namely from the heat of the fire by which I am sitting. Nothing is more obvious than the judgment that this thing is sending its likeness rather than something else into me.

I will now see whether these reasons are powerful enough. When I say here "I have been so taught by nature," all I have in mind is that I am driven by a spontaneous impulse to believe this, and not that some light of nature is showing me that it is true. These are two very different things. For whatever is shown me by this light of nature, for example, that from the fact that I doubt, it follows that I am, and the like, cannot in any way be doubtful. This is owing to the fact that there can be no other faculty that I can trust as much as this light and which could teach that these things are not true. But as far as natural impulses are concerned, in the *39* past I have often judged myself to have been driven by them to make the poorer choice when it was a question of choosing a good; and I fail to see why I should place any greater faith in them than in other matters.

Again, although these ideas do not depend upon my will, it does not follow that they necessarily proceed from things existing outside me. For just as these impulses about which I spoke just now seem to be different from my will, even though they are in me, so too perhaps there is also in me some other faculty, one not yet sufficiently known to me, which produces these ideas, just as it has always seemed up to now that ideas are formed in me without any help from external things when I am asleep.

And finally, even if these ideas did proceed from things other than myself, it does not therefore follow that they must resemble those things. Indeed it seems I have frequently noticed a vast difference in many respects. For example, I find within myself two distinct ideas of the sun. One idea is drawn, as it were, from the senses. Now it is this idea which, of all those that I take to be derived from outside me, is most in need of examination. By means of this idea the sun appears to me to be quite small. But there is another idea, one derived from astronomical reasoning, that is, it is elicited from certain notions that are innate in me, or else is fashioned by me in some other way. Through this idea the sun is shown to be several times larger than the earth. Both ideas surely cannot resemble the same sun existing outside me; and reason convinces me that the idea that seems to have emanated from the sun itself from so close is the very one that least resembles the sun.

All these points demonstrate sufficiently that up to this point it was not *40* a well-founded judgment but only a blind impulse that formed the basis of my belief that things existing outside me send ideas or images of themselves to me through the sense organs or by some other means.

But still another way occurs to me for inquiring whether some of the things of which there are ideas in me do exist outside me: insofar as these ideas are merely modes of thought, I see no inequality among them; they all seem to proceed from me in the same manner. But insofar as one idea represents one thing and another idea another thing, it is obvious that they do differ very greatly from one another. Unquestionably, those ideas that display substances to me are something more and, if I may say so, contain within themselves more objective reality than those which represent only modes or accidents. Again, the idea that enables me to understand a supreme deity, eternal, infinite, omniscient, omnipotent, and creator of all things other than himself, clearly has more objective reality within it than do those ideas through which finite substances are displayed.

Now it is indeed evident by the light of nature that there must be at least as much [reality] in the efficient and total cause as there is in the effect of that same cause. For whence, I ask, could an effect get its reality, if not from its cause? And how could the cause give that reality to the effect, unless it also possessed that reality? Hence it follows that something cannot come into being out of nothing, and also that what is more perfect *41* (that is, what contains in itself more reality) cannot come into being from what is less perfect. But this is manifestly true not merely for those effects whose reality is actual or formal, but also for ideas in which only objective reality is considered. For example, not only can a stone which did not exist previously not now begin to exist unless it is produced by something in which there is, either formally or eminently, everything that is in the stone; nor heat be introduced into a subject which was not already hot unless it is done by something that is of at least as perfect an order as heat—and the same for the rest—but it is also true that there can be in me no idea of heat, or of a stone, unless it is placed in me by some cause that has at least as much reality as I conceive to be in the heat or in the stone. For although this cause conveys none of its actual or formal reality to my idea, it should not be thought for that reason that it must be less real. Rather, the very nature of an idea is such that of itself it needs no formal reality other than what it borrows from my thought, of which it is a mode. But that a particular idea contains this as opposed to that objective reality is surely owing to some cause in which there is at least as much formal reality as there is objective reality contained in the idea. For if we assume that something is found in the idea that was not in its cause, then the idea gets that something from nothing. Yet as imperfect a mode of being as this is by which a thing exists in the intellect objectively through an idea, nevertheless it is plainly not nothing; hence it cannot get its being from nothing.

Moreover, even though the reality that I am considering in my ideas is merely objective reality, I ought not on that account to suspect that there *42* is no need for the same reality to be formally in the causes of these ideas, but that it suffices for it to be in them objectively. For just as the objective mode of being belongs to ideas by their very nature, so the formal mode of being belongs to the causes of ideas, at least to the first and preeminent ones, by their very nature. And although one idea can perhaps issue from another, nevertheless no infinite regress is permitted here; eventually some first idea must be reached whose cause is a sort of archetype that contains formally all the reality that is in the idea merely objectively. Thus it is clear to me by the light of nature that the ideas that are in me are like images that can easily fail to match the perfection of the things from which they have been drawn, but which can contain nothing greater or more perfect.

And the longer and more attentively I examine all these points, the more clearly and distinctly I know they are true. But what am I ultimately to conclude? If the objective reality of any of my ideas is found to be so great that I am certain that the same reality was not in me, either formally or eminently, and that therefore I myself cannot be the cause of the idea, then it necessarily follows that I am not alone in the world, but that something else, which is the cause of this idea, also exists. But if no such idea is found in me, I will have no argument whatsoever to make me certain of the existence of anything other than myself, for I have conscientiously reviewed all these arguments, and so far I have been unable to find any other.

Among my ideas, in addition to the one that displays me to myself (about which there can be no difficulty at this point), are others that *43* represent God, corporeal and inanimate things, angels, animals, and finally other men like myself.

As to the ideas that display other men, or animals, or angels, I easily understand that they could be fashioned from the ideas that I have of myself, of corporeal things, and of God—even if no men (except myself), no animals, and no angels existed in the world.

As to the ideas of corporeal things, there is nothing in them that is so great that it seems incapable of having originated from me. For if I investigate them thoroughly and examine each one individually in the way I examined the idea of wax yesterday, I notice that there are only a very few things in them that I perceive clearly and distinctly: namely, size, or extension in length, breadth, and depth; shape, which arises from the limits of this extension; position, which various things possessing shape have in relation to one another; and motion, or alteration in position. To these can be added substance, duration, and number. But as for the remaining items, such as light and colors, sounds, odors, tastes, heat and cold and other tactile qualities, I think of these only in a very confused and obscure manner, to the extent that I do not even know whether they are true or false, that is, whether the ideas I have of them are ideas of things or ideas of non-things. For although a short time ago I noted that falsity properly so called (or "formal" falsity) is to be found only in judgments, nevertheless there is another kind of falsity (called "material" falsity) which is found in ideas whenever they represent a non-thing as if it were a thing. For example, the ideas I have of heat and cold fall so far *44* short of being clear and distinct that I cannot tell from them whether cold is merely the privation of heat or whether heat is the privation of cold, or whether both are real qualities, or whether neither is. And because ideas can only be, as it were, of things, if it is true that cold is merely the absence of heat, then an idea that represents cold to me as something real and positive will not inappropriately be called false. The same holds for other similar ideas.

Assuredly I need not assign to these ideas an author distinct from myself. For if they were false, that is, if they were to represent non-things, I know by the light of nature that they proceed from nothing; that is, they are in me for no other reason than that something is lacking in my nature, and that my nature is not entirely perfect. If, on the other hand, these ideas are true, then because they exhibit so little reality to me that I cannot distinguish it from a non-thing, I see no reason why they cannot get their being from me.

As for what is clear and distinct in the ideas of corporeal things, it appears I could have borrowed some of these from the idea of myself: namely, substance, duration, number, and whatever else there may be of this type. For instance, I think that a stone is a substance, that is to say, a thing that is suitable for existing in itself; and likewise I think that I too am a substance. Despite the fact that I conceive myself to be a thinking thing and not an extended thing, whereas I conceive of a stone as an extended thing and not a thinking thing, and hence there is the greatest diversity between these two concepts, nevertheless they seem to agree with one another when considered under the rubric of substance. Furthermore, I perceive that I now exist and recall that I have previously existed for some time. And I have various thoughts and know how many of them
45 there are. It is in doing these things that I acquire the ideas of duration and number, which I can then apply to other things. However, none of the other components out of which the ideas of corporeal things are fashioned (namely extension, shape, position, and motion) are contained in me formally, since I am merely a thinking thing. But since these are only certain modes of a substance, whereas I am a substance, it seems possible that they are contained in me eminently.

Thus there remains only the idea of God. I must consider whether there is anything in this idea that could not have originated from me. I understand by the name "God" a certain substance that is infinite, independent, supremely intelligent and supremely powerful, and that created me along with everything else that exists—if anything else exists. Indeed all these are such that, the more carefully I focus my attention on them, the less possible it seems they could have arisen from myself alone. Thus, from what has been said, I must conclude that God necessarily exists.

For although the idea of substance is in me by virtue of the fact that I am a substance, that fact is not sufficient to explain my having the idea of an infinite substance, since I am finite, unless this idea proceeded from some substance which really was infinite.

Nor should I think that I do not perceive the infinite by means of a true idea, but only through a negation of the finite, just as I perceive rest and darkness by means of a negation of motion and light. On the contrary, I clearly understand that there is more reality in an infinite substance than there is in a finite one. Thus the perception of the infinite is somehow prior in me to the perception of the finite, that is, my perception of God is prior to my perception of myself. For how would I understand that I doubt and that I desire, that is, that I lack something and that I am not *46* wholly perfect, unless there were some idea in me of a more perfect being, by comparison with which I might recognize my defects?

Nor can it be said that this idea of God is perhaps materially false and thus can originate from nothing, as I remarked just now about the ideas of heat and cold, and the like. On the contrary, because it is the most clear and distinct and because it contains more objective reality than any other idea, no idea is in and of itself truer and has less of a basis for being suspected of falsehood. I maintain that this idea of a being that is supremely perfect and infinite is true in the highest degree. For although I could perhaps pretend that such a being does not exist, nevertheless I could not pretend that the idea of such a being discloses to me nothing real, as was the case with the idea of cold which I referred to earlier. It is indeed an idea that is utterly clear and distinct; for whatever I clearly and distinctly perceive to be real and true and to involve some perfection is wholly contained in that idea. It is no objection that I do not comprehend the infinite or that there are countless other things in God that I can in no way either comprehend or perhaps even touch with my thought. For the nature of the infinite is such that it is not comprehended by a being such as I, who am finite. And it is sufficient that I understand this very point and judge that all those things that I clearly perceive and that I know to contain some perfection—and perhaps even countless other things of which I am ignorant—are in God either formally or eminently. The result is that, of all the ideas that are in me, the idea that I have of God is the most true, the most clear and distinct.

But perhaps I am something greater than I myself understand. Perhaps all these perfections that I am attributing to God are somehow in me 47 potentially, although they do no yet assert themselves and are not yet actualized. For I now observe that my knowledge is gradually being increased, and I see nothing standing in the way of its being increased more and more to infinity. Moreover, I see no reason why, with my knowledge thus increased, I could not acquire all the remaining perfections of God. And, finally, if the potential for these perfections is in me already, I see no reason why this potential would not suffice to produce the idea of these perfections.

Yet none of these things can be the case. First, while it is true that my knowledge is gradually being increased and that there are many things in me potentially that are not yet actual, nevertheless, none of these pertains to the idea of God, in which there is nothing whatever that is potential. Indeed this gradual increase is itself a most certain proof of imperfection. Moreover, although my knowledge may always increase more and more, nevertheless I understand that this knowledge will never by this means be actually infinite, because it will never reach a point where it is incapable of greater increase. On the contrary, I judge God to be actually infinite, so that nothing can be added to his perfection. Finally, I perceive that the objective being of an idea cannot be produced by a merely potential being (which, strictly speaking, is nothing), but only by an actual or formal being.

Indeed there is nothing in all these things that is not manifest by the light of nature to one who is conscientious and attentive. But when I am less attentive, and the images of sensible things blind the mind's eye, I do not so easily recall why the idea of a being more perfect than me necessarily 48 proceeds from a being that really is more perfect. This being the case, it is appropriate to ask further whether I myself who have this idea could exist, if such a being did not exist.

From what source, then, do I derive my existence? Why, from myself, or from my parents, or from whatever other things there are that are less perfect than God. For nothing more perfect than God, or even as perfect as God, can be thought or imagined.

But if I got my being from myself, I would not doubt, nor would I desire, nor would I lack anything at all. For I would have given myself all the perfections of which I have some idea; in so doing, I myself would be God! I must not think that the things I lack could perhaps be more difficult to acquire than the ones I have now. On the contrary, it is obvious that it would have been much more difficult for me (that is, a thing or substance that thinks) to emerge out of nothing than it would be to acquire the knowledge of many things about which I am ignorant (these items of knowledge being merely accidents of that substance). Certainly, if I got this greater thing from myself, I would not have denied myself at least those things that can be had more easily. Nor would I have denied myself any of those other things that I perceive to be contained in the idea of God, for surely none of them seem to me more difficult to bring about. But if any of them were more difficult to bring about, they would certainly also seem more difficult to me, even if the remaining ones that I possess I got from myself, since it would be on account of them that I would experience that my power is limited.

Nor am I avoiding the force of these arguments, if I suppose that perhaps I have always existed as I do now, as if it then followed that no author of my existence need be sought. For because the entire span of one's life can be divided into countless parts, each one wholly independent *49* of the rest, it does not follow from the fact that I existed a short time ago that I must exist now, unless some cause, as it were, creates me all over again at this moment, that is to say, which preserves me. For it is obvious to one who pays close attention to the nature of time that plainly the same force and action are needed to preserve anything at each individual moment that it lasts as would be required to create that same thing anew, were it not yet in existence. Thus conservation differs from creation solely by virtue of a distinction of reason; this too is one of those things that are manifest by the light of nature.

Therefore I must now ask myself whether I possess some power by which I can bring it about that I myself, who now exist, will also exist a little later on. For since I am nothing but a thinking thing—or at least since I am now dealing simply and precisely with that part of me which is a thinking thing—if such a power were in me, then I would certainly be aware of it. But I observe that there is no such power; and from this very fact I know most clearly that I depend upon some being other than myself.

But perhaps this being is not God, and I have been produced either by my parents or by some other causes less perfect than God. On the contrary, as I said before, it is obvious that there must be at least as much in the cause as there is in the effect. Thus, regardless of what it is that eventually is assigned as my cause, because I am a thinking thing and have within me a certain idea of God, it must be granted that what caused me is also a thinking thing and it too has an idea of all the perfections which I attribute to God. And I can again inquire of this cause whether it got its existence from itself or from another cause. For if it got its existence from itself, it is evident from what has been said that it is itself God, because, having the power of existing in and of itself, it unquestionably also has the *50* power of actually possessing all the perfections of which it has in itself an idea—that is, all the perfections that I conceive to be in God. However, if it got its existence from another cause, I will once again inquire in similar fashion about this other cause: whether it got its existence from itself or from another cause, until finally I arrive at the ultimate cause, which will be God. For it is apparent enough that there can be no infinite regress here, especially since I am not dealing here merely with the cause that once produced me, but also and most especially with the cause that preserves me at the present time.

Nor can one fancy that perhaps several partial causes have concurred in bringing me into being, and that I have taken the ideas of the various perfections I attribute to God from a variety of causes, so that all of these perfections are found somewhere in the universe, but not all joined together in a single being—God. On the contrary, the unity, the simplicity, that is, the inseparability of all those features that are in God is one of the chief perfections that I understand to be in him. Certainly the idea of the unity of all his perfections could not have been placed in me by any cause from which I did not also get the ideas of the other perfections; for neither could some cause have made me understand them joined together and inseparable from one another, unless it also caused me to recognize what they were.

Finally, as to my parents, even if everything that I ever believed about them were true, still it is certainly not they who preserve me; nor is it they who in any way brought me into being, insofar as I am a thinking thing. Rather, they merely placed certain dispositions in the matter which I judged to contain me, that is, a mind, which now is the only thing I take *51* myself to be. And thus there can be no difficulty here concerning my parents. Indeed I have no choice but to conclude that the mere fact of my existing and of there being in me an idea of a most perfect being, that is, God, demonstrates most evidently that God too exists.

All that remains for me is to ask how I received this idea of God. For I did not draw it from the senses; it never came upon me unexpectedly, as is usually the case with the ideas of sensible things when these things present themselves (or seem to present themselves) to the external sense organs. Nor was it made by me, for I plainly can neither subtract anything from it nor add anything to it. Thus the only option remaining is that this idea is innate in me, just as the idea of myself is innate in me.

To be sure, it is not astonishing that in creating me, God should have endowed me with this idea, so that it would be like the mark of the craftsman impressed upon his work, although this mark need not be something distinct from the work itself. But the mere fact that God created me makes it highly plausible that I have somehow been made in his image and likeness, and that I perceive this likeness, in which the idea of God is contained, by means of the same faculty by which I perceive myself. That is, when I turn the mind's eye toward myself, I understand not only that I am something incomplete and dependent upon another, something aspiring indefinitely for greater and greater or better things, but also that the being on whom I depend has in himself all those greater things—not merely indefinitely and potentially, but infinitely and actually, and thus that he is God. The whole force of the argument rests on the fact that I recognize that it would be impossible for me to exist, being of such a *52* nature as I am (namely, having in me the idea of God), unless God did in fact exist. God, I say, that same being the idea of whom is in me: a being having all those perfections that I cannot comprehend, but can somehow touch with my thought, and a being subject to no defects whatever. From these considerations it is quite obvious that he cannot be a deceiver, for it is manifest by the light of nature that all fraud and deception depend on some defect.

But before examining this idea more closely and at the same time inquiring into other truths that can be gathered from it, at this point I want to spend some time contemplating this God, to ponder his attributes and, so far as the eye of my darkened mind can take me, to gaze upon, to admire, and to adore the beauty of this immense light. For just as we believe by faith that the greatest felicity of the next life consists solely in this contemplation of the divine majesty, so too we now experience that from the same contemplation, although it is much less perfect, the greatest pleasure of which we are capable in this life can be perceived.

MEDITATION FOUR: Concerning the True and the False

Lately I have become accustomed to withdrawing my mind from the senses, and I have carefully taken note of the fact that very few things are *53* truly perceived regarding corporeal things, although a great many more things are known regarding the human mind, and still many more things regarding God. The upshot is that I now have no difficulty directing my thought away from things that can be imagined to things that can be grasped only by the understanding and are wholly separate from matter. In fact the idea I clearly have of the human mind—insofar as it is a thinking thing, not extended in length, breadth or depth, and having nothing else from the body—is far more distinct than the idea of any corporeal thing.

And when I take note of the fact that I doubt, or that I am a thing that is incomplete and dependent, there comes to mind a clear and distinct idea of a being that is independent and complete, that is, an idea of God. And from the mere fact that such an idea is in me, or that I who have this idea exist, I draw the obvious conclusion that God also exists, and that my existence depends entirely upon him at each and every moment. This conclusion is so obvious that I am confident that the human mind can know nothing more evident or more certain. And now I seem to see a way by which I might progress from this contemplation of the true God, in whom, namely, are hidden all the treasures of the sciences and wisdom, to the knowledge of other things.

To begin with, I acknowledge that it is impossible for God ever to deceive me, for trickery or deception is always indicative of some imperfection. And although the ability to deceive seems to be an indication of cleverness or power, the will to deceive undoubtedly attests to maliciousness or weakness. Accordingly, deception is incompatible with God.

Next I experience that there is in me a certain faculty of judgment, *54* which, like everything else that is in me, I undoubtedly received from God. And since he does not wish to deceive me, he assuredly has not given me the sort of faculty with which I could ever make a mistake, when I use it properly.

No doubt regarding this matter would remain, but for the fact that it seems to follow from this that I am never capable of making a mistake. For if everything that is in me I got from God, and he gave me no faculty for making mistakes, it seems I am incapable of ever erring. And thus, so long as I think exclusively about God and focus my attention exclusively on him, I discern no cause of error or falsity. But once I turn my attention back on myself, I nevertheless experience that I am subject to countless errors. As I seek a cause of these errors, I notice that passing before me is not only a real and positive idea of God (that is, of a supremely perfect being), but also, as it were, a certain negative idea of nothingness (that is, of what is at the greatest possible distance from any perfection), and that I have been so constituted as a kind of middle ground between God and nothingness, or between the supreme being and non-being. Thus insofar as I have been created by the supreme being, there is nothing in me by means of which I might be deceived or be led into error; but insofar as I participate in nothingness or non-being, that is, insofar as I am not the supreme being and lack a great many things, it is not surprising that I make mistakes. Thus I certainly understand that error as such is not something real that depends upon God, but rather is merely a defect. And thus there is no need to account for my errors by positing a faculty given to me by God for this purpose. Rather, it just so happens that I make mistakes because the faculty of judging the truth, which I got from God, is not, in my case, infinite.

Still this is not yet altogether satisfactory; for error is not a pure negation, *55* but rather a privation or a lack of some knowledge that somehow ought to be in me. And when I attend to the nature of God, it seems impossible that he would have placed in me a faculty that is not perfect in its kind or that is lacking some perfection it ought to have. For if it is true that the more expert the craftsman, the more perfect the works he produces, what can that supreme creator of all things make that is not perfect in all respects? No doubt God could have created me such that I never erred. No doubt, again, God always wills what is best. Is it then better that I should be in error rather than not?

As I mull these things over more carefully, it occurs to me first that there is no reason to marvel at the fact that God should bring about certain things the reasons for which I do not understand. Nor is his existence therefore to be doubted because I happen to experience other things of which I fail to grasp why and how he made them. For since I know now that my nature is very weak and limited, whereas the nature of God is immense, incomprehensible, and infinite, this is sufficient for me also to know that he can make innumerable things whose causes escape me. For this reason alone the entire class of causes which people customarily derive from a thing's "end," I judge to be utterly useless in physics. It is not without rashness that I think myself capable of inquiring into the ends of God.

It also occurs to me that whenever we ask whether the works of God are perfect, we should keep in view not simply some one creature in isolation from the rest, but the universe as a whole. For perhaps something might rightfully appear very imperfect if it were all by itself, and yet be *56* most perfect, to the extent that it has the status of a part in the universe. And although subsequent to having decided to doubt everything, I have come to know with certainty only that I and God exist, nevertheless, after having taken note of the immense power of God, I cannot deny that many other things have been made by him, or at least could have been made by him. Thus I may have the status of a part in the universal scheme of things.

Next, as I focus more closely on myself and inquire into the nature of my errors (the only things that are indicative of some imperfection in me), I note that these errors depend on the simultaneous concurrence of two causes: the faculty of knowing that is in me and the faculty of choosing, that is, the free choice of the will, in other words, simultaneously on the intellect and will. Through the intellect alone I merely perceive ideas, about which I can render a judgment. Strictly speaking, no error is to be found in the intellect when properly viewed in this manner. For although perhaps there may exist countless things about which I have no idea, nevertheless it must not be said that, strictly speaking, I am deprived of these ideas but only that I lack them in a negative sense. This is because I cannot adduce an argument to prove that God ought to have given me a greater faculty of knowing than he did. No matter how expert a craftsman I understand him to be, still I do not for that reason believe he ought to have bestowed on each one of his works all the perfections that he can put into some. Nor, on the other hand, can I complain that the will or free choice I have received from God is insufficiently ample or perfect, since I experience that it is limited by no boundaries whatever. In fact, it seems *57* to be especially worth noting that no other things in me are so perfect or so great but that I understand that they can be still more perfect or greater. If, for example, I consider the faculty of understanding, I immediately recognize that in my case it is very small and quite limited, and at the very same time I form an idea of another much greater faculty of understanding—in fact, an understanding which is consummately great and infinite; and from the fact that I can form an idea of this faculty, I perceive that it pertains to the nature of God. Similarly, were I to examine the faculty of memory or imagination, or any of the other faculties, I would understand that in my case each of these is without exception feeble and limited, whereas in the case of God I understand each faculty to be boundless. It is only the will or free choice that I experience to be so great in me that I cannot grasp the idea of any greater faculty. This is so much the case that the will is the chief basis for my understanding that I bear a certain image and likeness of God. For although the faculty of willing is incomparably greater in God than it is in me, both by virtue of the knowledge and power that are joined to it and that render it more resolute and efficacious and by virtue of its object inasmuch as the divine will stretches over a greater number of things, nevertheless, when viewed in itself formally and precisely, God's faculty of willing does not appear to be any greater. This is owing to the fact that willing is merely a matter of being able to do or not do the same thing, that is, of being able to affirm or deny, to pursue or to shun; or better still, the will consists solely in the fact that when something is proposed to us by our intellect either to affirm or deny, to pursue or to shun, we are moved in such a way that we sense that we are determined to it by no external force. In order to be free I need not be capable of being moved in each direction; on the contrary, the more I am inclined toward one direction—either because I clearly understand that there is in it an aspect of the good and the true, or because God has thus *58* disposed the inner recesses of my thought—the more freely do I choose that direction. Nor indeed does divine grace or natural knowledge ever diminish one's freedom; rather, they increase and strengthen it. However, the indifference that I experience when there is no reason moving me more in one direction than in another is the lowest grade of freedom; it is indicative not of any perfection in freedom, but rather of a defect, that is, a certain negation in knowledge. Were I always to see clearly what is true and good, I would never deliberate about what is to be judged or chosen. In that event, although I would be entirely free, I could never be indifferent.

But from these considerations I perceive that the power of willing, which I got from God, is not, taken by itself, the cause of my errors, for it is most ample as well as perfect in its kind. Nor is my power of understanding the cause of my errors. For since I got my power of understanding from God, whatever I understand I doubtless understand rightly, and it is impossible for me to be deceived in this. What then is the source of my errors? They are owing simply to the fact that, since the will extends further than the intellect, I do not contain the will within the same boundaries; rather, I also extend it to things I do not understand. Because the will is indifferent in regard to such matters, it easily turns away from the true and the good; and in this way I am deceived and I sin.

For example, during these last few days I was examining whether anything in the world exists, and I noticed that, from the very fact that I was making this examination, it obviously followed that I exist. Nevertheless, I could not help judging that what I understood so clearly was true; not that I was coerced into making this judgment because of some external force, *59* but because a great light in my intellect gave way to a great inclination in my will, and the less indifferent I was, the more spontaneously and freely did I believe it. But now, in addition to my knowing that I exist, insofar as I am a certain thinking thing, I also observe a certain idea of corporeal nature. It happens that I am in doubt as to whether the thinking nature which is in me, or rather which I am, is something different from this corporeal nature, or whether both natures are one and the same thing. And I assume that as yet no consideration has occurred to my intellect to convince me of the one alternative rather than the other. Certainly in virtue of this very fact I am indifferent about whether to affirm or to deny either alternative, or even whether to make no judgment at all in the matter.

Moreover, this indifference extends not merely to things about which the intellect knows absolutely nothing, but extends generally to everything of which the intellect does not have a clear enough knowledge at the very time when the will is deliberating on them. For although probable guesses may pull me in one direction, the mere knowledge that they are only guesses and not certain and indubitable proofs is all it takes to push my assent in the opposite direction. These last few days have provided me with ample experience on this point. For all the beliefs that I had once held to be most true I have supposed to be utterly false, and for the sole reason that I determined that I could somehow raise doubts about them.

But if I hold off from making a judgment when I do not perceive what is true with sufficient clarity and distinctness, it is clear that I am acting properly and am not committing an error. But if instead I were to make *60* an assertion or a denial, then I am not using my freedom properly. Were I to select the alternative that is false, then obviously I will be in error. But were I to embrace the other alternative, it will be by sheer luck that I happen upon the truth; but I will still not be without fault, for it is manifest by the light of nature that a perception on the part of the intellect must always precede a determination on the part of the will. Inherent in this incorrect use of free will is the privation that constitutes the very essence of error: the privation, I say, present in this operation insofar as the operation proceeds from me, but not in the faculty given to me by God, nor even in its operation insofar as it depends upon him.

Indeed I have no cause for complaint on the grounds that God has not given me a greater power of understanding or a greater light of nature than he has, for it is of the essence of a finite intellect not to understand many things, and it is of the essence of a created intellect to be finite. Actually, instead of thinking that he has withheld from me or deprived me of those things that he has not given me, I ought to thank God, who never owed me anything, for what he has bestowed upon me.

Again, I have no cause for complaint on the grounds that God has given me a will that has a wider scope than my intellect. For since the will consists of merely one thing, something indivisible, as it were, it does not seem that its nature could withstand anything being removed from it. Indeed, the more ample the will is, the more I ought to thank the one who gave it to me.

Finally, I should not complain because God concurs with me in eliciting those acts of the will, that is those judgments, in which I am mistaken.

For insofar as those acts depend on God, they are absolutely true and good; and in a certain sense, there is greater perfection in me in being able to elicit those acts than in not being able to do so. But privation, in which alone the defining characteristic of falsehood and wrong-doing is *61* to be found, has no need whatever for God's concurrence, since a privation is not a thing, nor, when it is related to God as its cause, is it to be called a privation, but simply a negation. For it is surely no imperfection in God that he has given me the freedom to give or withhold my assent in those instances where he has not placed a clear and distinct perception in my intellect. But surely it is an imperfection in me that I do not use my freedom well and that I make judgments about things I do not properly understand. Nevertheless, I see that God could easily have brought it about that, while still being free and having finite knowledge, I should nonetheless never make a mistake. This result could have been achieved either by his endowing my intellect with a clear and distinct perception of everything about which I would ever deliberate, or by simply impressing the following rule so firmly upon my memory that I could never forget it: I should never judge anything that I do not clearly and distinctly understand. I readily understand that, considered as a totality, I would have been more perfect than I am now, had God made me that way. But I cannot therefore deny that it may somehow be a greater perfection in the universe as a whole that some of its parts are not immune to error, while others are, than if all of them were exactly alike. And I have no right to complain that the part God has wished me to play is not the principal and most perfect one of all.

Furthermore, even if I cannot abstain from errors in the first way mentioned above, which depends upon a clear perception of everything about which I must deliberate, nevertheless I can avoid error in the other way, which depends solely on my remembering to abstain from making *62* judgments whenever the truth of a given matter is not apparent. For although I experience a certain infirmity in myself, namely that I am unable to keep my attention constantly focused on one and the same item of knowledge, nevertheless, by attentive and often repeated meditation, I can bring it about that I call this rule to mind whenever the situation calls for it, and thus I would acquire a certain habit of not erring.

Since herein lies the greatest and chief perfection of man, I think today's meditation, in which I investigated the cause of error and falsity, was quite profitable. Nor can this cause be anything other than the one I have described; for as often as I restrain my will when I make judgments, so that it extends only to those matters that the intellect clearly and distinctly discloses to it, it plainly cannot happen that I err. For every clear and distinct perception is surely something, and hence it cannot come from nothing. On the contrary, it must necessarily have God for its author: God, I say, that supremely perfect being to whom it is repugnant to be a deceiver. Therefore the perception is most assuredly true. Today I have learned not merely what I must avoid so as never to make a mistake, but at the same time what I must do to attain truth. For I will indeed attain it, if only I pay enough attention to all the things that I perfectly understand, and separate them off from the rest, which I apprehend more confusedly and more obscurely. I will be conscientious about this in the future.

63 MEDITATION FIVE: Concerning the Essence of Material Things, and Again Concerning God, That He Exists

Several matters remain for me to examine concerning the attributes of God and myself, that is, concerning the nature of my mind. But perhaps I will take these up at some other time. For now, since I have noted what to avoid and what to do in order to attain the truth, nothing seems more pressing than that I try to free myself from the doubts into which I fell a few days ago, and that I see whether anything certain is to be had concerning material things.

Yet, before inquiring whether any such things exist outside me, I surely ought to consider the ideas of these things, insofar as they exist in my thought, and see which ones are distinct and which ones are confused.

I do indeed distinctly imagine the quantity that philosophers commonly call "continuous," that is, the extension of this quantity, or rather of the thing quantified in length, breadth and depth. I enumerate the various parts in it. I ascribe to these parts any sizes, shapes, positions, and local movements whatever; to these movements I ascribe any durations whatever.

Not only are these things manifestly known and transparent to me, viewed thus in a general way, but also, when I focus my attention on them, I perceive countless particulars concerning shapes, number, movement, *64* and the like. Their truth is so open and so much in accord with my nature that, when I first discover them, it seems I am not so much learning something new as recalling something I knew beforehand. In other words, it seems as though I am noticing things for the first time that were in fact in me for a long while, although I had not previously directed a mental gaze upon them.

What I believe must be considered above all here is the fact that I find within me countless ideas of certain things, that, even if perhaps they do not exist anywhere outside me, still cannot be said to be nothing. And although, in a sense, I think them at will, nevertheless they are not something I have fabricated; rather they have their own true and immutable natures. For example, when I imagine a triangle, even if perhaps no such figure exists outside my thought anywhere in the world and never has, the triangle still has a certain determinate nature, essence, or form which is unchangeable and eternal, which I did not fabricate, and which does not depend on my mind. This is evident from the fact that various properties can be demonstrated regarding this triangle: namely, that its three angles are equal to two right angles, that its longest side is opposite its largest angle, and so on. These are properties I now clearly acknowledge, whether I want to or not, even if I previously had given them no thought whatever when I imagined the triangle. For this reason, then, they were not fabricated by me.

It is irrelevant for me to say that perhaps the idea of a triangle came to me from external things through the sense organs because of course I have on occasion seen triangle-shaped bodies. For I can think of countless other figures, concerning which there can be no suspicion of their ever having entered me through the senses, and yet I can demonstrate various *65* properties of these figures, no less than I can those of the triangle. All these properties are patently true because I know them clearly, and thus they are something and not merely nothing. For it is obvious that whatever is true is something, and I have already demonstrated at some length that all that I know clearly is true. And even if I had not demonstrated this, certainly the nature of my mind is such that nevertheless I cannot refrain from assenting to these things, at least while I perceive them clearly. And I recall that even before now, when I used to keep my attention glued to the objects of the senses, I always took the truths I clearly recognized regarding figures, numbers, or other things pertaining to arithmetic, geometry or, in general, to pure and abstract mathematics to be the most certain of all.

But if, from the mere fact that I can bring forth from my thought the idea of something, it follows that all that I clearly and distinctly perceive to belong to that thing really does belong to it, then cannot this too be a basis for an argument proving the existence of God? Clearly the idea of God, that is, the idea of a supremely perfect being, is one I discover to be no less within me than the idea of any figure or number. And that it belongs to God's nature that he always exists is something I understand no less clearly and distinctly than is the case when I demonstrate in regard to some figure or number that something also belongs to the nature of that figure or number. Thus, even if not everything that I have meditated upon during these last few days were true, still the existence of God 66 ought to have for me at least the same degree of certainty that truths of mathematics had until now.

However, this point is not wholly obvious at first glance, but has a certain look of a sophism about it. Since in all other matters I have become accustomed to distinguishing existence from essence, I easily convince myself that it can even be separated from God's essence, and hence that God can be thought of as not existing. But nevertheless, it is obvious to anyone who pays close attention that existence can no more be separated from God's essence than its having three angles equal to two right angles can be separated from the essence of a triangle, or than that the idea of a valley can be separated from the idea of a mountain. Thus it is no less[1] contradictory to think of God (that is, a supremely perfect being) lacking existence (that is, lacking some perfection) than it is to think of a mountain without a valley.

But granted I can no more think of God as not existing than I can think of a mountain without a valley, nevertheless it surely does not follow from the fact that I think of a mountain without a valley that a mountain exists in the world. Likewise, from the fact that I think of God as existing, it does not seem to follow that God exists, for my thought imposes no necessity on things. And just as one may imagine a winged horse, without there being a horse that has wings, in the same way perhaps I can attach existence to God, even though no God exists.

But there is a sophism lurking here. From the fact that I am unable to 67 think of a mountain without a valley, it does not follow that a mountain or a valley exists anywhere, but only that, whether they exist or not, a mountain and a valley are inseparable from one another. But from the fact that I cannot think of God except as existing, it follows that existence is inseparable from God, and that for this reason he really exists. Not that my thought brings this about or imposes any necessity on anything; but rather the necessity of the thing itself, namely of the existence of God, forces me to think this. For I am not free to think of God without existence, that is, a supremely perfect being without a supreme perfection, as I am to imagine a horse with or without wings.

Further, it should not be said here that even though I surely need to

1. A literal translation of the Latin text (*non magis*) is "no more." This is obviously a misstatement on Descartes's part, since it contradicts his own clearly stated views.

assent to the existence of God once I have asserted that God has all perfections and that existence is one of these perfections, nevertheless that earlier assertion need not have been made. Likewise, I need not believe that all four-sided figures can be inscribed in a circle; but given that I posit this, it would then be necessary for me to admit that a rhombus can be inscribed in a circle. Yet this is obviously false. For although it is not necessary that I should ever happen upon any thought of God, nevertheless whenever I am of a mind to think of a being that is first and supreme, and bring forth the idea of God as it were from the storehouse of my mind, I must of necessity ascribe all perfections to him, even if I do not at that time enumerate them all or take notice of each one individually. This necessity plainly suffices so that afterwards, when I realize that existence is a perfection, I rightly conclude that a first and supreme being exists. In the same way, there is no necessity for me ever to imagine a triangle, but whenever I do wish to consider a rectilinear figure having but three angles, I must ascribe to it those properties on the basis of which *68* one rightly infers that the three angles of this figure are no greater than two right angles, even though I do not take note of this at the time. But when I inquire as to the figures that may be inscribed in a circle, there is absolutely no need whatever for my thinking that all four-sided figures are of this sort; for that matter, I cannot even fabricate such a thing, so long as I am of a mind to admit only what I clearly and distinctly understand. Consequently, there is a great difference between false assumptions of this sort and the true ideas that are inborn in me, the first and chief of which is the idea of God. For there are a great many ways in which I understand that this idea is not an invention that is dependent upon my thought, but is an image of a true and immutable nature. First, I cannot think of anything aside from God alone to whose essence existence belongs. Next, I cannot understand how there could be two or more Gods of this kind. Again, once I have asserted that one God now exists, I plainly see that it is necessary that he has existed from eternity and will endure for eternity. Finally, I perceive many other features in God, none of which I can remove or change.

But, whatever type of argument I use, it always comes down to the fact that the only things that fully convince me are those that I clearly and distinctly perceive. And although some of these things I thus perceive are obvious to everyone, while others are discovered only by those who look more closely and inquire carefully, nevertheless, once they have been discovered, they are considered no less certain than the others. For example, in the case of a right triangle, although it is not so readily apparent

69 that the square of the hypotenuse is equal to the sum of the squares of the other two sides as it is that the hypotenuse is opposite the largest angle, nevertheless, once the former has been ascertained, it is no less believed. However, as far as God is concerned, if I were not overwhelmed by prejudices and if the images of sensible things were not besieging my thought from all directions, I would certainly acknowledge nothing sooner or more easily than him. For what, in and of itself, is more manifest than that a supreme being exists, that is, that God, to whose essence alone existence belongs, exists?

And although I needed to pay close attention in order to perceive this, nevertheless I now am just as certain about this as I am about everything else that seems most certain. Moreover, I observe also that certitude about other things is so dependent on this, that without it nothing can ever be perfectly known.

For I am indeed of such a nature that, while I perceive something very clearly and distinctly, I cannot help believing it to be true. Nevertheless, my nature is also such that I cannot focus my mental gaze always on the same thing, so as to perceive it clearly. Often the memory of a previously made judgment may return when I am no longer attending to the arguments on account of which I made such a judgment. Thus, other arguments can be brought forward that would easily make me change my opinion, were I ignorant of God. And thus I would never have true and certain knowledge about anything, but merely fickle and changeable opinions. Thus, for example, when I consider the nature of a triangle, it appears most evident to me, steeped as I am in the principles of geometry, *70* that its three angles are equal to two right angles. And so long as I attend to its demonstration I cannot help believing this to be true. But no sooner do I turn the mind's eye away from the demonstration, than, however much I still recall that I had observed it most clearly, nevertheless, it can easily happen that I entertain doubts about whether it is true, were I ignorant of God. For I can convince myself that I have been so constituted by nature that I might occasionally be mistaken about those things I believe I perceive most evidently, especially when I recall that I have often taken many things to be true and certain, which other arguments have subsequently led me to judge to be false.

But once I perceived that there is a God, and also understood at the same time that everything else depends on him, and that he is not a deceiver, I then concluded that everything that I clearly and distinctly perceive is necessarily true. Hence even if I no longer attend to the reasons leading me to judge this to be true, so long as I merely recall that I did clearly and distinctly observe it, no counter-argument can be brought forward that might force me to doubt it. On the contrary, I have a true and certain knowledge of it. And not just of this one fact, but of everything else that I recall once having demonstrated, as in geometry, and so on. For what objections can now be raised against me? That I have been made such that I am often mistaken? But I now know that I cannot be mistaken in matters I plainly understand. That I have taken many things to be true and certain which subsequently I recognized to be false? But none of these were things I clearly and distinctly perceived. But I was ignorant of this rule for determining the truth, and I believed these things perhaps for other reasons which I later discovered were less firm. What then remains to be said? That perhaps I am dreaming, as I recently objected against myself, in other words, that everything I am now thinking of is no truer than what occurs to someone who is asleep? Be that as it may, this changes nothing; for certainly, even if I were dreaming, if anything is evident to *71* my intellect, then it is entirely true.

And thus I see plainly that the certainty and truth of every science depends exclusively upon the knowledge of the true God, to the extent that, prior to my becoming aware of him, I was incapable of achieving perfect knowledge about anything else. But now it is possible for me to achieve full and certain knowledge about countless things, both about God and other intellectual matters, as well as about the entirety of that corporeal nature which is the object of pure mathematics.

MEDITATION SIX: Concerning the Existence of Material Things, and the Real Distinction between Mind and Body

It remains for me to examine whether material things exist. Indeed I now know that they can exist, at least insofar as they are the object of pure mathematics, since I clearly and distinctly perceive them. For no doubt God is capable of bringing about everything that I am capable of perceiving in this way. And I have never judged that God was incapable of something, except when it was incompatible with my perceiving it distinctly. Moreover, from the faculty of imagination, which I notice I use while dealing with material things, it seems to follow that they exist. For to anyone paying very close attention to what imagination is, it appears to be simply a certain *72* application of the knowing faculty to a body intimately present to it, and which therefore exists.

To make this clear, I first examine the difference between imagination and pure intellection. So, for example, when I imagine a triangle, I not only understand that it is a figure bounded by three lines, but at the same time I also envisage with the mind's eye those lines as if they were present; and this is what I call "imagining." On the other hand, if I want to think about a chiliagon, I certainly understand that it is a figure consisting of a thousand sides, just as well as I understand that a triangle is a figure consisting of three sides, yet I do not imagine those thousand sides in the same way, or envisage them as if they were present. And although in that case—because of force of habit I always imagine something whenever I think about a corporeal thing—I may perchance represent to myself some figure in a confused fashion, nevertheless this figure is obviously not a chiliagon. For this figure is really no different from the figure I would represent to myself, were I thinking of a myriagon or any other figure with a large number of sides. Nor is this figure of any help in knowing the properties that differentiate a chiliagon from other polygons. But if the figure in question is a pentagon, I surely can understand its figure, just as was the case with the chiliagon, without the help of my imagination. But I can also imagine a pentagon by turning the mind's eye both to its five sides and at the same time to the area bounded by those sides. At this *73* point I am manifestly aware that I am in need of a peculiar sort of effort on the part of the mind in order to imagine, one that I do not employ in order to understand. This new effort on the part of the mind clearly shows the difference between imagination and pure intellection.

Moreover, I consider that this power of imagining that is in me, insofar as it differs from the power of understanding, is not required for my own essence, that is, the essence of my mind. For were I to be lacking this power, I would nevertheless undoubtedly remain the same entity I am now. Thus it seems to follow that the power of imagining depends upon something distinct from me. And I readily understand that, were a body to exist to which a mind is so joined that it may apply itself in order, as it were, to look at it any time it wishes, it could happen that it is by means of this very body that I imagine corporeal things. As a result, this mode of thinking may differ from pure intellection only in the sense that the mind, when it understands, in a sense turns toward itself and looks at one of the ideas that are in it; whereas when it imagines, it turns toward the body, and intuits in the body something that conforms to an idea either understood by the mind or perceived by sense. To be sure, I easily understand that the imagination can be actualized in this way, provided a body does exist. And since I can think of no other way of explaining imagination that is equally appropriate, I make a probable conjecture from this that a body exists. But this is only a probability. And even though I may examine everything carefully, nevertheless I do not yet see how the distinct idea of corporeal nature that I find in my imagination can enable me to develop an argument which necessarily concludes that some body exists.

But I am in the habit of imagining many other things, over and above that corporeal nature which is the object of pure mathematics, such as colors, sounds, tastes, pain, and the like, though not so distinctly. And I perceive these things better by means of the senses, from which, with the aid of the memory, they seem to have arrived at the imagination. Thus I should pay the same degree of attention to the senses, so that I might deal with them more appropriately. I must see whether I can obtain any reliable argument for the existence of corporeal things from those things that are perceived by the mode of thinking that I call "sense." *74*

First of all, to be sure, I will review here all the things I previously believed to be true because I had perceived them by means of the senses and the causes I had for thinking this. Next I will assess the causes why I later called them into doubt. Finally, I will consider what I must now believe about these things.

So first, I sensed that I had a head, hands, feet, and other members that comprised this body which I viewed as part of me, or perhaps even as the whole of me. I sensed that this body was found among many other bodies, by which my body can be affected in various beneficial or harmful ways. I gauged what was opportune by means of a certain sensation of pleasure, and what was inopportune by a sensation of pain. In addition to pain and pleasure, I also sensed within me hunger, thirst, and other such appetites, as well as certain bodily tendencies toward mirth, sadness, anger, and other such affects. And externally, besides the extension, shapes, and motions of bodies, I also sensed their hardness, heat, and other tactile qualities. I also sensed light, colors, odors, tastes, and sounds, on the basis of whose variety I distinguished the sky, the earth, the seas, and the other bodies, one from the other. Now given the ideas of all these qualities that presented themselves to my thought, and which were all that I properly and immediately sensed, still it was surely not without reason that I thought I sensed things that were manifestly different from my thought, namely, the bodies from which these ideas proceeded. For I knew by experience that these ideas came upon me utterly without my consent, to the extent that, wish as I may, I could not sense any object unless it was present to a sense organ. Nor could I fail to sense it when it was present. And since the ideas perceived by sense were much more vivid and explicit and even, in their own way, more distinct than any of those that I deliberately and knowingly formed through meditation or that I found impressed on my *75* memory, it seemed impossible that they came from myself. Thus the remaining alternative was that they came from other things. Since I had no knowledge of such things except from those same ideas themselves, I could not help entertaining the thought that they were similar to those ideas. Moreover, I also recalled that the use of the senses antedated the use of reason. And since I saw that the ideas that I myself fashioned were not as explicit as those that I perceived through the faculty of sense, and were for the most part composed of parts of the latter, I easily convinced myself that I had absolutely no idea in the intellect that I did not have beforehand in the sense faculty. Not without reason did I judge that this *76* body, which by a certain special right I called "mine," belongs more to me than did any other. For I could never be separated from it in the same way I could be from other bodies. I sensed all appetites and feelings in and on behalf of it. Finally, I noticed pain and pleasurable excitement in its parts, but not in other bodies external to it. But why should a certain sadness of spirit arise from some sensation or other of pain, and why should a certain elation arise from a sensation of excitement, or why should that peculiar twitching in the stomach, which I call hunger, warn me to have something to eat, or why should dryness in the throat warn me to take something to drink, and so on? I plainly had no explanation other than that I had been taught this way by nature. For there is no affinity whatsoever, at least none I am aware of, between this twitching in the stomach and the will to have something to eat, or between the sensation of something causing pain and the thought of sadness arising from this sensation. But nature also seems to have taught me everything else as well that I judged concerning the objects of the senses, for I had already convinced myself that this was how things were, prior to my assessing any of the arguments that might prove it.

Afterwards, however, many experiences gradually weakened any faith that I had in the senses. Towers that had seemed round from afar occasionally appeared square at close quarters. Very large statues mounted on their pedestals did not seem large to someone looking at them from ground level. And in countless other such instances I determined that judgments in matters of the external senses were in error. And not just the external *77* senses, but the internal senses as well. For what can be more intimate than pain? But I had sometimes heard it said by people whose leg or arm had been amputated that it seemed to them that they still occasionally sensed pain in the very limb they had lost. Thus, even in my own case it did not seem to be entirely certain that some bodily member was causing me pain, even though I did sense pain in it. To these causes for doubt I recently added two quite general ones. The first was that everything I ever thought I sensed while awake I could believe I also sometimes sensed while asleep, and since I do not believe that what I seem to sense in my dreams comes to me from things external to me, I saw no reason why I should hold this belief about those things I seem to be sensing while awake. The second was that, since I was still ignorant of the author of my origin (or at least pretended to be ignorant of it), I saw nothing to prevent my having been so constituted by nature that I should be mistaken even about what seemed to me most true. As to the arguments that used to convince me of the truth of sensible things, I found no difficulty responding to them. For since I seemed driven by nature toward many things about which reason tried to dissuade me, I did not think that what I was taught by nature deserved much credence. And even though the perceptions of the senses did not depend on my will, I did not think that we must therefore conclude that they came from things distinct from me, since perhaps there is some faculty in me, as yet unknown to me, that produces these perceptions.

But now, having begun to have a better knowledge of myself and the author of my origin, I am of the opinion that I must not rashly admit everything that I seem to derive from the senses; but neither, for that *78* matter, should I call everything into doubt.

First, I know that all the things that I clearly and distinctly understand can be made by God such as I understand them. For this reason, my ability clearly and distinctly to understand one thing without another suffices to make me certain that the one thing is different from the other, since they can be separated from each other, at least by God. The question as to the sort of power that might effect such a separation is not relevant to their being thought to be different. For this reason, from the fact that I know that I exist, and that at the same time I judge that obviously nothing else belongs to my nature or essence except that I am a thinking thing, I rightly conclude that my essence consists entirely in my being a thinking thing. And although perhaps (or rather, as I shall soon say, assuredly) I have a body that is very closely joined to me, nevertheless, because on the one hand I have a clear and distinct idea of myself, insofar as I am merely a thinking thing and not an extended thing, and because on the other hand I have a distinct idea of a body, insofar as it is merely an extended thing and not a thinking thing, it is certain that I am really distinct from my body, and can exist without it.

Moreover, I find in myself faculties for certain special modes of thinking, namely the faculties of imagining and sensing. I can clearly and distinctly understand myself in my entirety without these faculties, but not vice versa: I cannot understand them clearly and distinctly without me, that is, without a substance endowed with understanding in which they inhere, for they include an act of understanding in their formal concept. Thus I perceive them to be distinguished from me as modes from a thing. I also acknowledge that there are certain other faculties, such as those of moving *79* from one place to another, of taking on various shapes, and so on, that, like sensing or imagining, cannot be understood apart from some substance in which they inhere, and hence without which they cannot exist. But it is clear that these faculties, if in fact they exist, must be in a corporeal or extended substance, not in a substance endowed with understanding. For some extension is contained in a clear and distinct concept of them, though certainly not any understanding. Now there clearly is in me a passive faculty of sensing, that is, a faculty for receiving and knowing the ideas of sensible things; but I could not use it unless there also existed, either in me or in something else, a certain active faculty of producing or bringing about these ideas. But this faculty surely cannot be in me, since it clearly presupposes no act of understanding, and these ideas are produced without my cooperation and often even against my will. Therefore the only alternative is that it is in some substance different from me, containing either formally or eminently all the reality that exists objectively in the ideas produced by that faculty, as I have just noted above. Hence this substance is either a body, that is, a corporeal nature, which contains formally all that is contained objectively in the ideas, or else it is God, or some other creature more noble than a body, which contains eminently all that is contained objectively in the ideas. But since God is not a deceiver, it is patently obvious that he does not send me these ideas either immediately by himself, or even through the mediation of some creature that contains the objective reality of these ideas not formally but only eminently. For since God has given me no faculty whatsoever for making *80* this determination, but instead has given me a great inclination to believe that these ideas issue from corporeal things, I fail to see how God could be understood not to be a deceiver, if these ideas were to issue from a source other than corporeal things. And consequently corporeal things exist. Nevertheless, perhaps not all bodies exist exactly as I grasp them by sense, since this sensory grasp is in many cases very obscure and confused. But at least they do contain everything I clearly and distinctly understand—that is, everything, considered in a general sense, that is encompassed in the object of pure mathematics.

As far as the remaining matters are concerned, which are either merely particular (for example, that the sun is of such and such a size or shape, and so on) or less clearly understood (for example, light, sound, pain, and the like), even though these matters are very doubtful and uncertain, nevertheless the fact that God is no deceiver (and thus no falsity can be found in my opinions, unless there is also in me a faculty given me by God for the purpose of rectifying this falsity) offers me a definite hope of reaching the truth even in these matters. And surely there is no doubt that all that I am taught by nature has some truth to it; for by "nature," taken generally, I understand nothing other than God himself or the ordered network of created things which was instituted by God. By my own particular nature I understand nothing other than the combination of all the things bestowed upon me by God.

There is nothing that this nature teaches me more explicitly than that I have a body that is ill-disposed when I feel pain, that needs food and drink when I suffer hunger or thirst, and the like. Therefore, I should not doubt that there is some truth in this.

By means of these sensations of pain, hunger, thirst and so on, nature *81* also teaches not merely that I am present to my body in the way a sailor is present in a ship, but that I am most tightly joined and, so to speak, commingled with it, so much so that I and the body constitute one single thing. For if this were not the case, then I, who am only a thinking thing, would not sense pain when the body is injured; rather, I would perceive the wound by means of the pure intellect, just as a sailor perceives by sight whether anything in his ship is broken. And when the body is in need of food or drink, I should understand this explicitly, instead of having confused sensations of hunger and thirst. For clearly these sensations of thirst, hunger, pain, and so on are nothing but certain confused modes of thinking arising from the union and, as it were, the commingling of the mind with the body.

Moreover, I am also taught by nature that various other bodies exist around my body, some of which are to be pursued, while others are to be avoided. And to be sure, from the fact that I sense a wide variety of colors, sounds, odors, tastes, levels of heat, and grades of roughness, and the like, I rightly conclude that in the bodies from which these different perceptions of the senses proceed there are differences corresponding to the different perceptions—though perhaps the latter do not resemble the former. And from the fact that some of these perceptions are pleasant while others are unpleasant, it is plainly certain that my body, or rather my whole self, insofar as I am comprised of a body and a mind, can be affected by various beneficial and harmful bodies in the vicinity.

82 Granted, there are many other things that I seem to have been taught by nature; nevertheless it was not really nature that taught them to me but a certain habit of making reckless judgments. And thus it could easily happen that these judgments are false: for example, that any space where there is absolutely nothing happening to move my senses is empty; or that there is something in a hot body that bears an exact likeness to the idea of heat that is in me; or that in a white or green body there is the same whiteness or greenness that I sense; or that in a bitter or sweet body there is the same taste, and so on; or that stars and towers and any other distant bodies have the same size and shape that they present to my senses, and other things of this sort. But to ensure that my perceptions in this matter are sufficiently distinct, I ought to define more precisely what exactly I mean when I say that I am "taught something by nature." For I am taking "nature" here more narrowly than the combination of everything bestowed on me by God. For this combination embraces many things that belong exclusively to my mind, such as my perceiving that what has been done cannot be undone, and everything else that is known by the light of nature. That is not what I am talking about here. There are also many things that belong exclusively to the body, such as that it tends to move downward, and so on. I am not dealing with these either, but only with what God has bestowed on me insofar as I am comprised of mind and body. Accordingly, it is this nature that teaches me to avoid things that produce a sensation of pain and to pursue things that produce a sensation of pleasure, and the like. But it does not appear that nature teaches us to conclude anything, besides these things, from these sense perceptions unless the intellect has first conducted its own inquiry regarding things external to us. For it 83 seems to belong exclusively to the mind, and not to the composite of mind and body, to know the truth in these matters. Thus, although a star affects my eye no more than does the flame from a small torch, still there is no real or positive tendency in my eye toward believing that the star is no larger than the flame. Yet, ever since my youth, I have made this judgment without any reason for doing so. And although I feel heat as I draw closer to the fire, and I also feel pain upon drawing too close to it, there is not a single argument that persuades me that there is something in the fire similar to that heat, any more than to that pain. On the contrary, I am convinced only that there is something in the fire that, regardless of what it finally turns out to be, causes in us those sensations of heat or pain. And although there may be nothing in a given space that moves the senses, it does not therefore follow that there is no body in it. But I see that in these and many other instances I have been in the habit of subverting the order

of nature. For admittedly I use the perceptions of the senses (which are properly given by nature only for signifying to the mind what things are useful or harmful to the composite of which it is a part, and to that extent they are clear and distinct enough) as reliable rules for immediately discerning what is the essence of bodies located outside us. Yet they signify nothing about that except quite obscurely and confusedly.

I have already examined in sufficient detail how it could happen that my judgments are false, despite the goodness of God. But a new difficulty now arises regarding those very things that nature shows me are either to be sought out or avoided, as well as the internal sensations where I seem to have detected errors, as for example, when someone is deluded by a food's pleasant taste to eat the poison hidden inside it. In this case, *84* however, he is driven by nature only toward desiring the thing in which the pleasurable taste is found, but not toward the poison, of which he obviously is unaware. I can only conclude that this nature is not omniscient. This is not remarkable, since man is a limited thing, and thus only what is of limited perfection befits him.

But we not infrequently err even in those things to which nature impels us. Take, for example, the case of those who are ill and who desire food or drink that will soon afterwards be injurious to them. Perhaps it could be said here that they erred because their nature was corrupt. However, this does not remove our difficulty, for a sick man is no less a creature of God than a healthy one, and thus it seems no less inconsistent that the sick man got a deception-prone nature from God. And a clock made of wheels and counter-weights follows all the laws of nature no less closely when it has been badly constructed and does not tell time accurately than it does when it completely satisfies the wish of its maker. Likewise, I might regard a man's body as a kind of mechanism that is outfitted with and composed of bones, nerves, muscles, veins, blood and skin in such a way that, even if no mind existed in it, the man's body would still exhibit all the same motions that are in it now except for those motions that proceed either from a command of the will or, consequently, from the mind. I easily recognize that it would be natural for this body, were it, say, suffering from dropsy and experiencing dryness in the throat (which typically produces a thirst sensation in the mind), and also so disposed by its nerves and other parts to take something to drink, the result of which would be to exacerbate the illness. This is as natural as for a body without any such illness to be moved by the same dryness in the throat to take something *85* to drink that is useful to it. And given the intended purpose of the clock, I could say that it deviates from its nature when it fails to tell the right time. And similarly, considering the mechanism of the human body in terms of its being equipped for the motions that typically occur in it, I may think that it too is deviating from its nature, if its throat were dry when having something to drink is not beneficial to its conservation. Nevertheless, I am well aware that this last use of "nature" differs greatly from the other. For this latter "nature" is merely a designation dependent on my thought, since it compares a man in poor health and a poorly constructed clock with the ideas of a healthy man and of a well-made clock, a designation extrinsic to the things to which it is applied. But by "nature" taken in the former sense, I understand something that is really in things, and thus is not without some truth.

When we say, then, in the case of the body suffering from dropsy, that its "nature" is corrupt, given the fact that it has a parched throat and yet does not need something to drink, "nature" obviously is merely an extrinsic designation. Nevertheless, in the case of the composite, that is, of a mind joined to such a body, it is not a pure designation, but a true error of nature that this body should be thirsty when having something to drink would be harmful to it. It therefore remains to inquire here how the goodness of God does not prevent "nature," thus considered, from being deceptive.

Now my first observation here is that there is a great difference between a mind and a body in that a body, by its very nature, is always divisible.
86 On the other hand, the mind is utterly indivisible. For when I consider the mind, that is, myself insofar as I am only a thinking thing, I cannot distinguish any parts within me; rather, I understand myself to be manifestly one complete thing. Although the entire mind seems to be united to the entire body, nevertheless, were a foot or an arm or any other bodily part to be amputated, I know that nothing has been taken away from the mind on that account. Nor can the faculties of willing, sensing, understanding, and so on be called "parts" of the mind, since it is one and the same mind that wills, senses, and understands. On the other hand, there is no corporeal or extended thing I can think of that I may not in my thought easily divide into parts; and in this way I understand that it is divisible. This consideration alone would suffice to teach me that the mind is wholly diverse from the body, had I not yet known it well enough in any other way.

My second observation is that my mind is not immediately affected by all the parts of the body, but only by the brain, or perhaps even by just one small part of the brain, namely, by that part where the "common" sense is said to reside. Whenever this part of the brain is disposed in the same manner, it presents the same thing to the mind, even if the other parts of the body are able meanwhile to be related in diverse ways. Countless experiments show this, none of which need be reviewed here.

My next observation is that the nature of the body is such that whenever any of its parts can be moved by another part some distance away, it can also be moved in the same manner by any of the parts that lie between them, even if this more distant part is doing nothing. For example, in the cord ABCD, if the final part D is pulled, the first part A would be moved *87* in exactly the same manner as it could be, if one of the intermediate parts B or C were pulled, while the end part D remained immobile. Likewise, when I feel a pain in my foot, physics teaches me that this sensation took place by means of nerves distributed throughout the foot, like stretched cords extending from the foot all the way to the brain. When these nerves are pulled in the foot, they also pull on the inner parts of the brain to which they extend, and produce a certain motion in them. This motion has been constituted by nature so as to affect the mind with a sensation of pain, as if it occurred in the foot. But because these nerves need to pass through the shin, thigh, loins, back, and neck to get from the foot to the brain, it can happen that even if it is not the part in the foot but merely one of the intermediate parts that is being struck, the very same movement will occur in the brain that would occur were the foot badly injured. The inevitable result will be that the mind feels the same pain. The same opinion should hold for any other sensation.

My final observation is that, since any given motion occurring in that part of the brain immediately affecting the mind produces but one sensation in it, I can think of no better arrangement than that it produces the one sensation that, of all the ones it is able to produce, is most especially and most often conducive to the maintenance of a healthy man. Moreover, experience shows that all the sensations bestowed on us by nature are like this. Hence there is absolutely nothing to be found in them that does not bear witness to God's power and goodness. Thus, for example, when the *88* nerves in the foot are agitated in a violent and unusual manner, this motion of theirs extends through the marrow of the spine to the inner reaches of the brain, where it gives the mind the sign to sense something, namely, the pain as if it is occurring in the foot. This provokes the mind to do its utmost to move away from the cause of the pain, since it is seen as harmful to the foot. But the nature of man could have been so constituted by God that this same motion in the brain might have indicated something else to the mind: for example, either the motion itself as it occurs in the brain, or in the foot, or in some place in between, or something else entirely different. But nothing else would have served so well the maintenance of the body. Similarly, when we need something to drink, a certain dryness arises in the throat that moves the nerves in the throat, and, by means of them, the inner parts of the brain. And this motion affects the mind with a sensation of thirst, because in this entire affair nothing is more useful for us to know than that we need something to drink in order to maintain our health; the same holds in the other cases.

From these considerations it is utterly apparent that, notwithstanding the immense goodness of God, the nature of man, insofar as it is composed of mind and body, cannot help being sometimes mistaken. For if some cause, not in the foot but in some other part through which the nerves extend from the foot to the brain, or perhaps even in the brain itself, were to produce the same motion that would normally be produced by a badly injured foot, the pain will be felt as if it were in the foot, and the senses will naturally be deceived. For since an identical motion in the brain can only bring about an identical sensation in the mind, and it is more frequently the case that this motion is wont to arise on account of a cause that harms the foot than on account of some other thing existing elsewhere, *89* it is reasonable that the motion should always show pain to the mind as something belonging to the foot rather than to some other part. And if dryness in the throat does not arise, as is normal, because taking something to drink contributes to bodily health, but from a contrary cause, as happens in the case of someone with dropsy, then it is far better that it should deceive on that occasion than that it should always be deceptive when the body is in good health. The same holds for the other cases.

This consideration is most helpful, not only for my noticing all the errors to which my nature is liable, but also for enabling me to correct or avoid them without difficulty. To be sure, I know that all the senses set forth what is true more frequently than what is false regarding what concerns the welfare of the body. Moreover, I can nearly always make use of several of them in order to examine the same thing. Furthermore, I can use my memory, which connects current happenings with past ones, and my intellect, which now has examined all the causes of error. Hence I should no longer fear that those things that are daily shown me by the senses are false. On the contrary, the hyperbolic doubts of the last few days ought to be rejected as ludicrous. This goes especially for the chief reason for doubting, which dealt with my failure to distinguish being asleep from being awake. For I now notice that there is a considerable difference between these two; dreams are never joined by the memory with all the other actions of life, as is the case with those actions that occur when one is awake. For surely, if, while I am awake, someone were suddenly to appear to me and then immediately disappear, as occurs in dreams, so that I see neither where he came from nor where he went, it is not without reason that I would judge him to be a ghost or a phantom conjured up in *90* my brain, rather than a true man. But when these things happen, and I notice distinctly where they come from, where they are now, and when they come to me, and when I connect my perception of them without interruption with the whole rest of my life, I am clearly certain that these perceptions have happened to me not while I was dreaming but while I was awake. Nor ought I have even the least doubt regarding the truth of these things, if, having mustered all the senses, in addition to my memory and my intellect, in order to examine them, nothing is passed on to me by one of these sources that conflicts with the others. For from the fact that God is no deceiver, it follows that I am in no way mistaken in these matters. But because the need to get things done does not always permit us the leisure for such a careful inquiry, we must confess that the life of man is apt to commit errors regarding particular things, and we must acknowledge the infirmity of our nature.